互联网+职业技能系列

职业入门 | 基础知识 | 系统进阶 | 专项提高

Java EE 企业级
框架开发实战教程

Spring Boot+Shiro+JPA | 微课版

Java EE Enterprise Framework Development

蜗牛学院 胡平 陈良

人民邮电出版社

北京

图书在版编目（CIP）数据

JavaEE企业级框架开发实战教程：Spring Boot+Shiro+JPA：微课版 / 蜗牛学院，胡平，陈良编著. -- 北京：人民邮电出版社，2020.6
互联网+职业技能系列
ISBN 978-7-115-52261-0

Ⅰ. ①J… Ⅱ. ①蜗… ②胡… ③陈… Ⅲ. ①JAVA语言—程序设计—教材 Ⅳ. ①TP312.8

中国版本图书馆CIP数据核字(2019)第226630号

内 容 提 要

本书较为全面地介绍了企业级开发框架的构成及应用。全书共 9 章，以 JDK 1.8 为基础，详细地介绍了从传统 Java Web 开发到企业级框架开发的演变发展，其知识点涵盖 Servlet、JDBC、Spring、Spring MVC、MyBatis、Shiro、Redis、缓存，以及脚手架工具 Spring Boot 和 Spring Data 等；以案例驱动的方式对知识点进行讲解，并在第 9 章通过贯穿案例讲解项目从研发到上线的过程，以练习和操作实践，帮助读者巩固所学内容。

本书可以作为高校计算机相关专业的教材，也可以作为有一定 Java 基础的计算机爱好者的自学用书。

◆ 编　著　蜗牛学院　胡　平　陈　良
　　责任编辑　左仲海
　　责任印制　王　郁　马振武

◆ 人民邮电出版社出版发行　北京市丰台区成寿寺路11号
　　邮编　100164　电子邮件　315@ptpress.com.cn
　　网址　https://www.ptpress.com.cn
　　三河市君旺印务有限公司印刷

◆ 开本：787×1092　1/16
　　印张：15.5　　　　　　　2020年6月第1版
　　字数：428千字　　　　　2020年6月河北第1次印刷

定价：49.80 元

读者服务热线：(010)81055256　印装质量热线：(010)81055316
反盗版热线：(010)81055315
广告经营许可证：京东工商广登字20170147号

前言
Foreword

  Java 技术的发展已有二十多年，相关的开发技术、组件、框架已非常成熟与丰富，相关应用的数量也日益增多，市场占有率不断提高，Java 已成为广大软件开发从业人员必须掌握的关键技术之一。目前，市面上很多关于企业级应用框架的书籍讲解只侧重某一个或几个框架，对现有的企业级应用框架讲解不全面，导致最终的实战案例也只针对部分框架实现，不太满足目前应用开发的需求。本书从传统的应用开发到企业级应用框架开发，讲解了目前市面上所涉及的大部分框架，并引入了最新的 Spring Boot、Spring Data，最后以实战项目向读者展示了整个项目的设计、实现及部署。

  本书编者有着多年的实际项目开发经验和丰富的教学经验。本书由浅入深地带领读者从零基础快速掌握企业级应用框架开发，并使读者能将理论结合实际，应用各种技术解决开发中的问题，完成业务需求。

  本书的主要特点如下。

1. 内容丰富、实用

  本书定位于企业级框架开发实战，内容主要包括传统应用开发（Servlet+JDBC）、Spring、Spring MVC、MyBatis、Redis 等常用的开发框架及缓存实现方式，同时讲解了 Shiro 框架来提升应用的安全性，并结合市场分析新增了 Spring Boot、Spring Data 等框架来简化应用的开发。

2. 实际项目开发与理论教学紧密结合

  为了使读者能够快速地掌握相关技术并将其熟练地运用在实际项目中，本书根据实际项目设计了大量的相关实训，第 9 章设计的项目可供读者进行独立学习与训练。

3. 组织合理、有效

  本书按照由浅入深的顺序，在逐渐丰富系统功能的同时，引入了相关技术与知识，实现技术讲解与训练的二合一，着重培养读者的动手能力。

  为方便读者使用，图书配套的教学资源均免费赠送给读者，读者可登录人邮教育社区（www.ryjiaoyu.com）或蜗牛学院官网（www.woniuxy.com）下载。读者也可加入蜗牛学院 IT 技术交流群，QQ 群号码为 934213545。

  由于编者水平有限，加之时间仓促，书中疏漏和不足之处在所难免，殷切希望广大读者批评指正。编者 E-mail：137273278@qq.com 或 13258413979@163.com。

编　者
2020 年 4 月

目录 Contents

第 1 章　传统 Java Web 开发　1

1.1　Servlet 概述　2
　1.1.1　了解 Servlet　2
　1.1.2　动态页面 JSP　6
　1.1.3　过滤器　19
1.2　Servlet 项目实战　23
　1.2.1　开发环境搭建　23
　1.2.2　MySQL 数据库搭建　23
　1.2.3　Servlet 请求处理　28
　1.2.4　Filter 权限控制　40

第 2 章　JavaEE 框架开发——SSM　44

2.1　MyBatis 概述　45
　2.1.1　了解 MyBatis　45
　2.1.2　MyBatis 数据持久化　46
　2.1.3　MyBatis 动态代理开发　51
　2.1.4　MyBatis 关系映射　54
2.2　Spring 概述　58
　2.2.1　了解 Spring　58
　2.2.2　Spring 的 IoC 容器　60
　2.2.3　Spring 的 AOP 编程　62
2.3　Spring MVC 概述　66
　2.3.1　Spring MVC 简介　66
　2.3.2　Spring MVC 请求处理　67
　2.3.3　注解开发　71
2.4　整合开发　73
　2.4.1　搭建 Spring 开发环境　73
　2.4.2　Spring 集成 MyBatis　76
　2.4.3　Spring 集成 Spring MVC　77
2.5　开发实战　78
　2.5.1　项目简介　78
　2.5.2　开发思路　79
　2.5.3　代码实现　79

第 3 章　Spring Boot　82

3.1　Spring Boot 概述　83
　3.1.1　了解 Spring Boot　83
　3.1.2　Spring Boot 的核心功能　83
　3.1.3　Spring Boot 示例　84
3.2　Spring Boot 核心　87
　3.2.1　自动配置　87
　3.2.2　自定义 starter　92

第 4 章　Spring Data　95

4.1　数据持久化　96
　4.1.1　了解数据持久化　96
　4.1.2　常用的数据持久化技术　96
4.2　持久化实现　96
　4.2.1　关系型数据库的持久化实现　96
　4.2.2　非关系型数据库的持久化实现　100
4.3　Spring Data　101
　4.3.1　Spring Data 入门　101
　4.3.2　Spring Data JPA　102
　4.3.3　Spring Data Redis　109

第 5 章　模板引擎　112

5.1　常用模板引擎　113
　5.1.1　模板引擎　113
　5.1.2　Spring Boot 对模板引擎的支持　114
5.2　FreeMarker 引擎　114
　5.2.1　了解 FreeMarker　114
　5.2.2　FreeMarker 类型　117
　5.2.3　FreeMarker 模板　118

第 6 章　Shiro 权限管理　122

6.1　Shiro 简介　123
6.2　用户认证　124
6.3　用户授权　127
6.4　Realm　130
6.5　基于 Shiro 的 Web 开发　136

第 7 章　Redis　142

7.1　认识 Redis　143
   7.1.1　RDBMS 与 NoSQL　143
   7.1.2　Redis 安装　144
   7.1.3　Redis 命令　146
7.2　Jedis 访问 Redis　152
   7.2.1　常用 API　152
   7.2.2　Spring 与 Jedis 的集成　161

第 8 章　缓存　162

8.1　缓存实现方案　163
8.2　Ehcache 实现　163
8.3　Redis 实现　174
8.4　其他缓存操作　177

第 9 章　项目实战　186

9.1　项目介绍　187
9.2　实战开发　187
   9.2.1　数据库设计　187
   9.2.2　环境搭建　190
   9.2.3　用户管理　194
   9.2.4　角色管理　211
   9.2.5　菜单管理　224
   9.2.6　权限控制　237
   9.2.7　项目部署　240

第1章

传统Java Web开发

本章导读

■ 本章主要介绍传统的 Web 开发应用，使读者对 Java Web 开发有一定的基础认识。同时，本章也为 Java EE 框架开发提供了基础知识。

学习目标

（1）充分理解Java动态网页技术Servlet。
（2）充分理解动态网页技术JSP。
（3）熟练运用EL表达式、JSTL标签与JSP页面。
（4）熟练运用过滤器进行登录、权限验证。

1.1 Servlet 概述

1.1.1 了解 Servlet

在互联网发展初期，浏览器访问到的页面通常是静态的 HTML 页面，即页面只包含了 HTML 文档的内容，与用户并没有任何交互。但是随着互联网技术的发展，静态网页编码的工作量大到几乎无法完成，动态网页技术随之出现。

动态网页不仅可以动态显示数据，还可以与用户做交互，如完成登录、注册等一系列动作。不同用户访问相同的网页时，经常会发现所显示的内容是不相同的，这样的网页就是经过服务器动态生成的。由于数据存储在数据库中，网页中显示的数据对应不同用户的数据库数据，因此，动态页面能够适用于大多数用户，程序员的编码工作量可大大减少，网页呈现的内容也愈加丰富多彩。

Servlet 就是一种动态网页技术。Servlet 是由 Java 语言编写的 Web 服务端程序，同其他 Java 程序一样，Servlet 的运行需要 JRE 或 JDK 的支持，同时，Servlet 程序还需要放置在支持 Servlet 的 Web 服务器中运行。目前市面上有很多支持 Servlet 的 Web 服务器，其中，Tomcat 是比较轻便且开源免费的服务器。

Servlet 以面向对象的形式解释了 HTTP 请求和响应内容，它将 HTTP 请求的内容封装为 HttpServletRequest 对象，将响应内容封装为 HttpServletResponse 对象。用户访问某个网址显示的网页内容，就是服务器通过 HttpServletResponse 对象发送到浏览器的数据。下面通过一个简单的例子为读者演示 Servlet 的开发流程。

首先，需要配置 Servlet 的开发环境，这里使用 Eclipse 来进行配置。

（1）在 Eclipse 中创建 Maven 项目。

（2）选择 "File" → "New" → "Other" 选项，在弹出的 "New Maven Project" 窗口中输入 "Maven"，选择 "Maven Project" 选项，单击 "Next" 按钮。

（3）勾选 "Create a simple project（skip archetype selection）" 复选框，如图 1-1 所示。

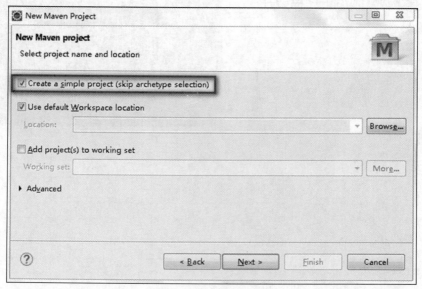

图 1-1　勾选 "Create a simple project（skip archetype selection）" 复选框

（4）单击 "Next" 按钮，填写 "Group Id" "Artifact Id"，这两个值是项目的 "坐标"。通常情况下，

"Group Id"填写的是公司域名，而"Artifact Id"填写的是项目名称。由于创建的是 Web 项目，因此需要设置打包方式"Packaging"为"war"，单击"Finish"按钮，如图 1-2 所示。

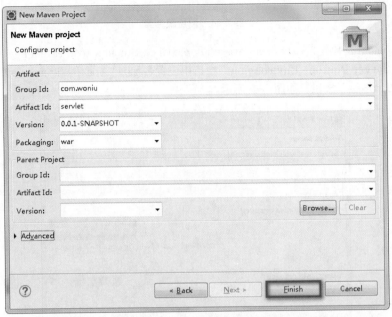

图 1-2　填写公司域名与项目名称，选择打包方式

（5）此时项目报错，这是因为 Eclipse 没有自动创建项目需要的 web.xml 配置文件，可以复制，也可以创建此文件。这里使用 Deployment 自动创建一份该配置文件，右键单击"Deployment Descriptor：servlet"选项，在弹出的快捷菜单中，选择"Generate Deployment Descriptor"选项，会自动创建项目需要的 web.xml 文件。最终，项目结构如图 1-3 所示。

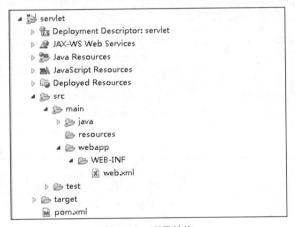

图 1-3　项目结构

项目创建完成后，需要为 Java Web 项目提供运行环境，这里使用 Tomcat，以下为 Tomcat 的配置步骤。

（1）下载 Tomcat 7.0。打开 Tomcat 官网，将 Tomcat 7.0 下载并解压到 D 盘中。

（2）配置 Server。选择"Window"→"Show View"→"Servers"选项，配置服务器，如图 1-4 所示。

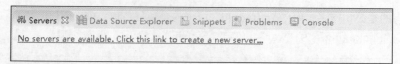

图 1-4 配置服务器

（3）选择 Tomcat。单击"No servers are available.Click this link to create a new server"超链接，在弹出的"New Server Runtime Environment"窗口中设置名称为"Apache Tomcat v7.0"，单击"Next"按钮，再单击"Browse"按钮，如图 1-5 所示。

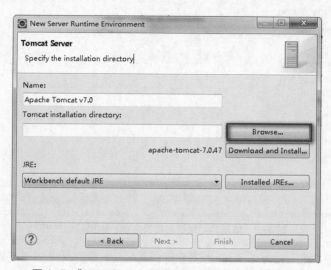

图 1-5 "New Server Runtime Environment"窗口

（4）选择刚才下载的 Tomcat 7.0 目录即可，单击"Finish"按钮，如图 1-6 所示。

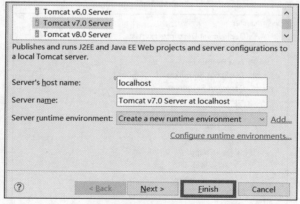

图 1-6 选择下载目录

（5）配置 Tomcat。在弹出的 Tomcat 配置对话框中找到"Servlet Locations"，选择"Use Tomcat installation"选项，按"Ctrl+S"快捷键保存。此时，通过 Eclipse 添加到 Tomcat 中的项目会部署在 Tomcat 根目录的 wtpwebapps 目录中。

（6）启动 Tomcat。右键单击"Tomcat"选项，在弹出的快捷菜单中选择"Start"选项，控制台打印启动信息，启动成功后打开浏览器并访问 http://localhost:8080，如进入图 1-7 所示的界面表示 Tomcat 配置成功。

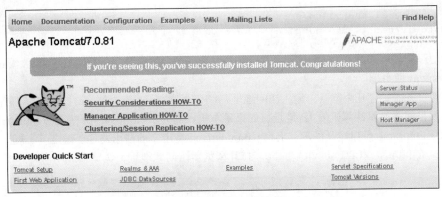

图 1-7　Tomcat 配置成功

Tomcat 配置成功之后，即可开始进行编码工作，一个 Web 项目的 Servlet 编码主要包括以下几个步骤。

（1）导入 JAR 包。使用 Maven 导入 Servlet 的 JAR 包，在 pom.xml 中添加依赖。

```xml
<dependencies>
  <!-- Servlet -->
  <dependency>
      <groupId>javax.servlet</groupId>
      <artifactId>servlet-api</artifactId>
      <version>2.5</version>
      <scope>provided</scope>
  </dependency>
</dependencies>
```

（2）编写 Servlet。在 src/java/main 中创建包 com.woniu.servlet，在包中创建类 HelloServlet，继承自 HttpServlet 类，并重写 service 方法（当然，也可以重写 doGet 或 doPost 方法）。

```java
public class HelloServlet extends HttpServlet{
    @Override
    //重写service方法
    public void service(HttpServletRequest request, HttpServletResponse response) throws ServletException, IOException {
        //配置UTF-8字符集以支持中文
        response.setContentType("text/html;charst=utf-8");
        response.setCharacterEncoding("UTF-8");
        //获取输出流以输出数据到浏览器中
        PrintWriter pw = response.getWriter();
        pw.write("我是Servlet服务端发送给浏览器的数据:Hello Servlet! ");
        pw.close();
    }
}
```

（3）配置请求路径。路径配置即指浏览器请求到达 Web 容器，并指定处理该请求的 Servlet 类。需要在 web.xml 中以标签的形式进行配置。完整的配置包括"servlet"标签与"servlet-mapping"标签，"servlet"标签中的"servlet-name"标签表示对 Servlet 进行命名，在整个 Web 容器中该值是唯一的，"servlet-class"标签需要填写完整的类路径。"servlet-mapping"是请求映射标签，表示 HTTP 请求与 Servlet 类的对应关系，对应关系通过"servlet-name"的值进行关联。

```xml
<servlet>
  <servlet-name>helloServlet</servlet-name>
  <servlet-class>com.woniu.servlet.HelloServlet</servlet-class>
</servlet>
<servlet-mapping>
```

```
    <servlet-name>helloServlet</servlet-name>
    <url-pattern>/hello</url-pattern>
</servlet-mapping>
```

（4）启动服务器。选择"Tomcat"→"Add and Remove"选项，将编写好的项目添加到 Tomcat 中，启动 Tomcat 服务器。

（5）访问服务器。打开浏览器，输入访问地址，URL 格式为 http://ip:port/appname/namespace。其中，ip 为项目部署所属计算机的 IP 地址，如果是本机，则为 localhost；port 为 Tomcat 访问端口，默认为 8080；appname 是自定义的项目名称；namespace 是定义的请求路径。根据以上规则，访问地址应当是 http://localhost:8080/servlet/hello。打开浏览器访问此地址，浏览器显示内容如图 1-8 所示。

图 1-8　浏览器显示内容

1.1.2　动态页面 JSP

通过 Servlet 可以实现页面的动态数据展示，在输出数据时，只需要通过 PrintWriter 对象动态地添加数据到内容中即可，也可以添加 HTML 标签和 CSS 样式，甚至可以使用 PrintWriter 对象向浏览器输出一个完整的 HTML 网页。下面通过代码来实现此功能。

V1-1　Tomcat 配置

```java
public class HelloServlet extends HttpServlet{
    @Override
    //重写service方法
    public void service(HttpServletRequest request, HttpServletResponse response) throws ServletException, IOException {
        response.setContentType("text/html;charst=utf-8");
        response.setCharacterEncoding("utf-8");
        String name = "张三";
        int age = 16;
        PrintWriter pw =  response.getWriter();
        //向浏览器输出数据，数据可以是一个完整的HTML文档
        pw.write("<p style='color:red'>你好，我叫"+name+",今年"+age+"岁！<p>");
        pw.close();
    }
}
```

动态的数据一般是通过数据库查询得到的，本案例中假设是从数据库查询得到的数据，页面显示内容如图 1-9 所示。

图 1-9　页面显示内容

V1-2　Servlet

可以看到，Servlet 可以实现动态数据展示，也可以为数据设置样式。但是，通常情况下，一个页面的内容非常多，页面样式也需要经过调试才能满足需求，在这种情况下，编码工作将变得非常复杂，基于此，JSP（Java Server Pages，Java 服务器页面）技术应运而生。

JSP 是由 Sun 公司（现已被甲骨文公司收购）联合其他公司建立的一种动态网页技术，是在 HTML

文件中嵌入 Java 代码的一种技术，JSP 文档的扩展名为.jsp。

下面在项目的 webapp 目录中编写文档。右键单击"webapp"选项，在弹出的快捷菜单中选择"New"→"JSP File"选项，创建 JSP 文档，如图 1-10 所示。

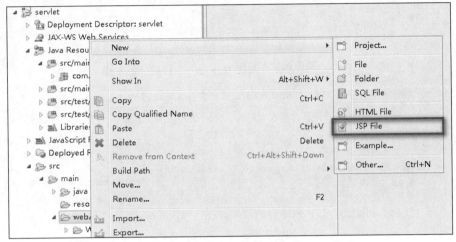

图 1-10　创建 JSP 文档

创建的 JSP 文档的顶部多了一段使用<%@%>包含的代码。

```
<%@ page language="java" contentType="text/html; charset=UTF-8"
    pageEncoding="UTF-8"%>
```

这是 JSP 文档的声明，pageEncoding 表示当前 JSP 文档的编码；contentType 表示服务器发送给客户端的 MIME 类型及编码。文档的"body"标签中添加了如下代码。

```
<body>
    Hello, MyServlet!!
</body>
```

启动服务器，访问路径 http://localhost:8080/servlet/hello.jsp，如图 1-11 所示。

图 1-11　访问 JSP 文档

一个 JSP 页面中包含以下 3 种 Java 代码。

（1）表达式。<%=表达式内容%>，可以是变量、变量的运算、有返回值的方法等。

（2）Java 片段。<%代码%>，可以出现在 Java 方法中的代码都可以作为片段。

（3）声明。<%!声明内容%>，用以声明成员变量、方法，代码内容和在 Java 中声明成员变量及成员方法的内容一致。

JSP 是属于服务端的文档，而属于客户端的浏览器可以解释 HTML、CSS、JavaScript 等文档，但无法解释 Java 语言，Java 语言只能由服务器进行解释。实际上，JSP 文档在服务端运行，会被 Web 容器解释为一个 Servlet 类。Web 容器对 JSP 页面进行解释时遵循以下规则。

（1）JSP 文档中的 HTML 代码内容会被翻译为 Servlet 类中的"out.write()"语句。

（2）JSP 文档中的表达式最终显示的是其运行的结果。

（3）JSP 文档中的 Java 片段会被翻译为 Servlet 类中 service 方法中的代码。

（4）JSP 文档中的声明会被翻译为 Servlet 类中的成员变量或成员方法。

下面对 JSP 页面中的 3 种 Java 代码进行展示。

```
<body>
    Hello,MyServlet!!<br>
    <!-- 声明:声明成员变量、成员方法 -->
    声明: <%!int i = 100;
        public String hello(){
            return "hello";
        }
    %>
    <br>
    <!-- 表达式:运行简单的算式、有返回值的方法，第一个算式调用的是声明中的变量i-->
    计算: i+100=<%=i+100 %>
    <br>
    调用hello(): <%=hello() %>
    <br>
    <!-- Java片段:Java片段会被翻译成Servlet 类中的Java代码 -->
    Java片段: <%out.println(i); %>
</body>
```

启动服务器，访问 hello.jsp，JSP 翻译后的效果如图 1-12 所示。

图 1-12　JSP 翻译后的效果

可以看到，在 JSP 页面中加入 Java 代码是可以正确执行的，这就是 JSP 的强大之处，可以将 JSP 理解为类 Servlet。

不知道读者有没有发现，前一个例子中的 hello.jsp 页面中使用到了 "out.println(i);" 这样一段代码，其中，i 是在前面声明的变量，但是 "out" 并没有进行声明，而在执行过程中，代码也没有报错、抛出异常，这是为什么呢？实际上，为了使用户编程更加方便，JSP 页面中内置了 9 个可以供用户使用的对象，"out" 就是其中之一。JSP 内置对象如表 1-1 所示。

表 1-1　JSP 内置对象

内置对象	说明
out	转译后对应 JspWriter 对象，其内部关联一个 PrintWriter 对象
request	转译后对应 HttpServletRequest/ServletRequest 对象
response	转译后对应 HttpServletResponse/ServletResponse 对象
config	转译后对应 ServletConfig 对象
application	转译后对应 ServletContext 对象
session	转译后对应 HttpSession 对象
pageContext	转译后对应 PageContext 对象，它提供了 JSP 页面资源的封装，并可设置页面范围属性
exception	转译后对应 Throwable 对象，代表由其他 JSP 页面抛出的异常对象，只会出现在 JSP 错误页面（isErrorPage 设置为 true 的 JSP 页面）中
page	转译后对应 this

在一个完整的 Web 项目中，通常是 JSP 页面和 Servlet 各司其职，Servlet 主要用于完成请求的逻辑处理，而 JSP 更多的是为逻辑处理之后的数据展示提供方便。在一个完整的请求过程中，既需要使用 Servlet 进行逻辑判断，又需要使用 JSP 进行数据展示，相当于是在多个 Servlet 之间进行跳转，这样的功能称为转发。

转发是指一个 Web 组件（Servlet/JSP）未能完成的处理通过容器转交给另一个 Web 组件完成。通常情况下，一个 Servlet 获取数据之后，将这些数据转发给 JSP，由这个 JSP 来展示这些数据。下面使用 Servlet 和 JSP 完成一个简单的登录案例。

（1）login.jsp 代码如下。

```
<body>
    <!--简单的登录页面 -->
    <form action="login" method="post">
        用户名：<input name="uname"><br>
        密    码：
        <input name="password" type="password"><br>
        <input type="submit" value="登录">
    </form>
</body>
```

（2）Servlet 代码如下。

```
public void service(HttpServletRequest req, HttpServletResponse resp) throws ServletException, IOException {
    //获取前端传入的uname属性的值
    String uname = req.getParameter("uname");
    //获取前端传入的password属性的值
    String password = req.getParameter("password");
    //如果用户名正确，则进行密码判断，这里假设正确的用户名为zhangsan、密码为123456
    if("zhangsan".equals(uname)) {
        if("123456".equals(password)) {
            //获取到index.jsp的转发器
            RequestDispatcher dispatcher =
                    req.getRequestDispatcher("index.jsp");
            //进行转发
            dispatcher.forward(req, resp);
            //因为转发之后的代码会继续执行，所以使用return结束方法
            return;
        }
    }
    //若用户名和密码不匹配，则转发到login.jsp
    RequestDispatcher dispatcher =
                    req.getRequestDispatcher("login.jsp");
    //进行转发
    dispatcher.forward(req, resp);
}
```

（3）web.xml 新增的配置代码如下。

```
<servlet>
    <servlet-name>login</servlet-name>
    <servlet-class>com.woniu.servlet.LoginServlet</servlet-class>
</servlet>
<servlet-mapping>
    <servlet-name>login</servlet-name>
    <url-pattern>/login</url-pattern>
</servlet-mapping>
```

（4）若输入错误的用户名和密码，则转发到 login.jsp 页面；若输入正确的用户名和密码，则转发到

index.jsp 页面（在实际项目中，通常是登录成功之后的欢迎页面）。

但是在实际使用网站登录功能的时候，如果输入错误的用户名或密码，则会有相应的提示；登录成功之后的欢迎页面中也会有相应的个人信息。使用 Servlet 和 JSP 可以完成这些简单功能，修改代码如下。

（1）login.jsp 代码如下。

```html
<body>
    <!-- 简单的登录页面 -->
    <form action="login" method="post">
        用户名：
            <input name="uname"><span style="color:red">
            <%=request.getAttribute("msg") %></span><br>
        密   码：<input name="password" type="password">
            <br>
        <input type="submit" value="登录">
    </form>
</body>
```

（2）Servlet 代码如下。

```java
@Override
    public void service(HttpServletRequest req, HttpServletResponse resp) throws ServletException, IOException {
        String uname = req.getParameter("uname");
        String password = req.getParameter("password");
        if("zhangsan".equals(uname)) {
            if("123456".equals(password)) {
                //登录成功，将用户数据存储在request对象中
                req.setAttribute("uname", uname);
                RequestDispatcher dispatcher = 
                        req.getRequestDispatcher("index.jsp");
                dispatcher.forward(req, resp);
                return;
            }
        }
        //给出用户提示，但是不用提示是密码错误还是用户名错误
        req.setAttribute("msg", "用户名或密码错误");
        RequestDispatcher dispatcher = 
                    req.getRequestDispatcher("login.jsp");
        dispatcher.forward(req, resp);
    }
```

（3）index.jsp 代码如下。

```
Welcome ,<%=request.getAttribute("uname") %>!
```

（4）用户名或密码输入错误时，显示效果如图 1-13 所示。

（5）用户名和密码输入正确时，显示效果如图 1-14 所示。

图 1-13　用户名或密码输入错误

图 1-14　用户名和密码输入正确

此案例就使用到了转发和 HttpServletRequest 对象的特性，转发的请求之间数据可以共享，数据

的共享就是通过 setAttribute(name,val) 方法进行设置的，而取出数据则是通过 getAttribute(name) 实现的。

转发还有一个特点，即浏览器地址不会发生改变。由于浏览器地址没有发生改变，所以会导致另外一个问题，即表单数据的重复提交，用户刷新登录之后的页面，浏览器会给出提示信息，询问是否重复提交表单数据，如图 1-15 所示。

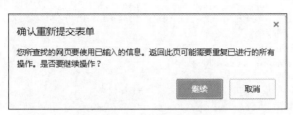

图 1-15 浏览器提示信息

表单数据的重复提交会引发一系列问题。例如，提交个人信息时，如果用户多次刷新页面，那么在数据库中会多次插入用户的数据，或者进行多次删除、更新等操作，这都会使数据出现异常。所以，表单数据的重复提交是需要避免的，在提交表单数据的地方，不能使用转发作为跳转页面的方式，而应使用 Servlet/JSP 提供的重定向（Redirect）作为跳转页面的机制。

重定向指用户发起请求处理完成之后，服务器返回 302 状态码及新的 URL 地址，浏览器会再次发起请求，访问服务器返回的 URL 地址。重定向相当于是浏览器发起了两次请求，而 request 对象的生命周期只能存在于当次请求，即之前使用 request 对象传递数据的方式不能再次使用。那么有没有一种作用范围比 request 对象更大的能够传递数据的对象呢？实际上，在 Web 应用中，JSP 创建的对象有一定的生命周期，也有可能被其他组件或对象访问。对象的生命周期和可访问性称为作用域。而 JSP 提供了 4 种作用域供编程者使用，如表 1-2 所示。

表 1-2 JSP 提供的 4 种作用域

隐式对象	说明	作用范围
page	转译后对应 JspWriter 对象，其内部关联一个 PrintWriter 对象	当前页面有效
request	转译后对应 HttpServletRequest/ServletRequest 对象	一个请求有效
session	转译后对应 HttpServletResponse/ServletResponse 对象	一次会话有效
application	转译后对应 ServletConfig 对象	整个项目有效

作用域范围由小到大为 page→request→session→application。转发只有一次请求，所以两个 Servlet 可以使用 request 对象传递数据，而重定向属于两个请求，不能再使用 request 对象传递数据。可以使用 session 或 application 对象传递数据，但是 application 的作用范围最大，通常存放的是一些公共资源数据，而这里的登录属于当前用户的信息，所以选择 session 来传递数据，Servlet 代码如下：

```
package com.woniu.servlet;

import java.io.IOException;

import javax.servlet.RequestDispatcher;
import javax.servlet.ServletException;
import javax.servlet.http.HttpServlet;
import javax.servlet.http.HttpServletRequest;
import javax.servlet.http.HttpServletResponse;
import javax.servlet.http.HttpSession;
```

```
public class LoginServlet extends HttpServlet{

    @Override
    public void service(HttpServletRequest req, HttpServletResponse resp) throws
ServletException, IOException {
        String uname = req.getParameter("uname");
        String password = req.getParameter("password");
        //通过request获取session对象
        HttpSession session = req.getSession();
        if("zhangsan".equals(uname)) {
            if("123456".equals(password)) {
                //登录成功,将用户数据存储在session对象中
                session.setAttribute("uname", uname);
                //重定向到index.jsp
                resp.sendRedirect("index.jsp");
                return;
            }
        }
        //给出用户提示,但是不用提示是密码错误还是用户名错误
        session.setAttribute("msg", "用户名或密码错误");
        resp.sendRedirect("login.jsp");
    }
}
```

login.jsp 代码如下。

```
<body>
    <!-- 简单的登录页面 -->
    <form action="login" method="post">
        用户名:
            <input name="uname"><span style="color:red">
            <%=request.getAttribute("msg") %></span><br>
        密    码: <input name="password" type="password">
            <br>
        <input type="submit" value="登录">
    </form>
</body>
```

登录成功的页面如图1-16所示。

图1-16 登录成功的页面

登录失败的页面如图1-17所示。

图1-17 登录失败的页面

可以看到，使用重定向的方式提交表单，浏览器地址会发生改变，此时刷新浏览器，不会再出现重复提交表单的提醒。在实际开发中，应当根据业务需求选择合适的跳转方式。

当 JSP 页面内容越来越多时，例如，在显示一个完整的列表信息时或数据的获取量大时，会发现 JSP 页面中会出现非常多的 Java 代码，这样虽然可以实现功能，但是总体 Java 代码多且复杂，又影响页面美观。下面为读者介绍一种简化 JSP 页面代码开发的功能，称为表达式语言（Expression Language，EL）。

EL 可使 JSP 写起来更加简单。EL 的灵感来自于 ECMAScript 和 XPath 表达式语言，它提供了在 JSP 中简化表达式的方法，使 JSP 的代码更加简练。

使用 EL 可以完成简单的计算，取出作用域中的值，以及取出 JavaBean 对象中的属性值，其基本语法为${exper}。下面为读者介绍 EL 的几种最为常用的语法。

（1）计算。在 webapp 目录中创建 el.jsp 文件，添加如下代码。

```jsp
<%@ page language="java" contentType="text/html; charset=UTF-8"
    pageEncoding="UTF-8"%>
<!DOCTYPE html PUBLIC "-//W3C//DTD HTML 4.01 Transitional//EN"
 "http://www.w3.org/TR/html4/loose.dtd">
<html>
<head>
<meta http-equiv="Content-Type" content="text/html; charset=UTF-8">
<title>EL表达式</title>
</head>
<body>
    EL表达式计算：<br>
    <!-- 加运算 -->
       1、${3+1} <br>
    <!-- 或运算 -->
       2、${true || false}<br>
    <!-- 三目运算 -->
       3、${1==1?"张三":"李四"}<br>
</body>
</html>
```

浏览器访问效果如图 1-18 所示。

EL 表达式可以完成大部分的运算功能，其他功能交由读者自己测试。

（2）取出前端传入的参数。在 el.jsp 中添加如下代码。

```jsp
EL表达式接收前端参数：<br>
    <!--${param.uname } 相当于request.getParameter("uname") -->
       ${param.uname }
```

访问 el.jsp，同时取出前端传入的参数，如图 1-19 所示。

图 1-18 浏览器访问效果

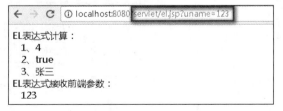

图 1-19 取出前端传入的参数

（3）获取 request 作用域传递的数据。创建 ParmServlet 类的代码如下。

```java
@Override
    protected void service(HttpServletRequest req, HttpServletResponse resp) throws ServletException, IOException {
```

```
    //绑定数据到request对象,数据以键/值对的形式存储、获取
    req.setAttribute("msg", "你好, 我是绑定在Request对象上的对象");
    //转发到el.jsp
    req.getRequestDispatcher("el.jsp").forward(req, resp);
}
```

在 web.xml 中添加如下代码。

```
<servlet>
  <servlet-name>parm</servlet-name>
  <servlet-class>com.woniu.servlet.ParmServlet</servlet-class>
</servlet>
<servlet-mapping>
  <servlet-name>parm</servlet-name>
  <url-pattern>/parm</url-pattern>
</servlet-mapping>
```

在 el.jsp 中添加如下代码。

```
EL表达式获取request作用域数据：<br>
<!--${requestScope.msg} 相当于 <%=session.getAttribute("msg") %>-->
       ${requestScope.msg }
```

访问请求 parm，访问结果如图 1-20 所示。

图 1-20　访问结果（1）

（4）获取 request 作用域数据的简单写法。在 el.jsp 中添加如下代码。

```
EL表达式获取request作用域数据简单写法：<br>
       ${msg }<br>
```

访问 parm 请求，访问结果如图 1-21 所示。

图 1-21　访问结果（2）

前面演示了两种获取 request 作用域数据的语法。实质上如果使用${name}语法，EL 表达式会依次从 4 种作用域中获取绑定的 name 数据，从左至右顺序依次为 page、request、session、application，从 page 开始，查找到数据则显示，没有查找到数据则继续从 request 对象中查找，依次运行；而${requestScope.name}则指定从 request 作用域中获取数据，获取到则显示，获取不到则不显示。作用域对象与 EL 表达式语法如表 1-3 所示。

表 1-3 作用域对象与 EL 表达式语法

作用域对象	EL 表达式语法
page	${pageScope.name}
request	${requestScope.name}
session	${sessionScope.name}
application	${applicationScope.name}

${name}语法效率偏低，但取值范围更大；${requestScope.name}语法效率高，但取值范围小。在确定取值范围的情况下，可以使用第二种语法获取数据；如果不确定取值范围，则推荐使用第一种语法。

（5）获取 JavaBean 对象的属性值。创建 com.woniu.entity 包，创建 User 类的代码如下。

```
package com.woniu.entity;

public class User {
    private String name;
    private int age;
    //getter setter...
}
```

创建 BeanServlet 类的代码如下。

```
protected void service(HttpServletRequest req, HttpServletResponse resp) throws
ServletException, IOException {
        //获取session对象传递数据或使用
        //request对象传递数据的效果都是一样的,这里主要为读者演示使用session
        //传递数据的效果及写法
        HttpSession session = req.getSession();
        User user = new User();
        user.setName("张三");
        user.setAge(20);
        //将创建的user对象绑定到session对象中
        session.setAttribute("user", user);
        //转发到el.jsp
        req.getRequestDispatcher("el.jsp").forward(req, resp);
    }
```

在 web.xml 中添加如下代码。

```
<servlet>
  <servlet-name>bean</servlet-name>
  <servlet-class>com.woniu.servlet.BeanServlet</servlet-class>
</servlet>
<servlet-mapping>
  <servlet-name>bean</servlet-name>
  <url-pattern>/bean</url-pattern>
</servlet-mapping>
```

在 el.jsp 中添加如下代码。

```
EL表达式获取JavaBean对象的属性值: <br>
<!--语法解释: 获取到作用域的user数据,再获取到user中的name属性值-->
   ${user.name }<br>
```

打开浏览器，访问请求 Bean，访问结果如图 1-22 所示。

图 1-22　访问结果（3）

（6）获取集合中的值。修改 BeanServlet，代码如下。

```
protected void service(HttpServletRequest req, HttpServletResponse resp) throws ServletException, IOException {
    HttpSession session = req.getSession();
    User user = new User();
    user.setName("张三");
    user.setAge(20);
    session.setAttribute("user", user);
    //创建Map对象，存入name和age属性
    Map<String,Object> map = new HashMap<String,Object>();
    map.put("name","张三");
    map.put("age", 25);
    session.setAttribute("map", map);
    //创建list对象，存入数据
    List<String> list = new ArrayList<String>();
    list.add("abc");
    list.add("xyz");
    session.setAttribute("list", list);
    //转发到el.jsp
    req.getRequestDispatcher("el.jsp").forward(req, resp);
}
```

在 el.jsp 中添加如下代码。

```
EL表达式获取集合的值：<br>
<!-- 获取map中的数据，通过键获取值，若键不存在则不显示 -->
   我叫${map.name},今年${map.age }岁了 ${map.sex }<br>
<!-- 获取list中的数据，通过下标获取，若下标不存在则不显示，不会抛出异常-->
${list[0] }、${list[1] }、${list[2]}<br>
```

访问请求 Bean，访问结果如图 1-23 所示。

图 1-23　访问结果（4）

前面介绍了 EL 表达式的使用方式。EL 表达式无法处理所有的页面操作，如列表显示、条件判断显示等，通常会使用 EL 表达式与 JSP 标准标签库（JSP Standard Tag Library，JSTL）结合的方式来完成这些功能。

JSTL 是一个实现 Web 应用程序中常见通用功能的定制标签库集。

使用 JSTL 需要将 JSTL 的 JAR 包引入 classpath，这里在 pom.xml 中添加依赖。

```xml
<!-- jstl -->
<dependency>
    <groupId>javax.servlet</groupId>
    <artifactId>jstl</artifactId>
    <version>1.2</version>
</dependency>
```

需要在 JSP 页面中引入所需的标签，在 webapp 目录中创建 jstl.jsp，并在文件开头添加如下代码。

```
<!--taglib表示引入标签库，      uri为标签库地址，    prefix为标签库别名，此名称可以在当前页面中使用-->
<%@ taglib uri="http://java.sun.com/jsp/jstl/core" prefix="c"%>
```

Core 标签库是 JSTL 的核心标签库，其中包含常用的处理逻辑。下面介绍 Core 中的常用标签。

（1）if 标签，用于做判断，类似于 Java 中的 if 程序结构。

① 创建 JstlServlet 类的代码如下。

```java
protected void service(HttpServletRequest req, HttpServletResponse resp) throws
ServletException, IOException {
    //绑定数据到request对象中
    req.setAttribute("name", "李四");
    req.setAttribute("age", 25);
    //获取转发器并转发到jstl.jsp
    req.getRequestDispatcher("jstl.jsp").forward(req, resp);
}
```

② 在 jstl.jsp 中添加如下代码。

```
${name }是一个
  <!-- test表示一个boolean值，var指定了一个变量，用于存储test的计算结果，此
       变量可以通过EL表达式获取，以下代码就是通过这种方式构造一个if else 结构
   -->
  <c:if test="${age>25 }" var="is">
      年轻人
  </c:if>
  <c:if test="${!is }">
      中年人
  </c:if>
```

③ 在 web.xml 中添加如下代码。

```xml
<servlet>
    <servlet-name>jstl</servlet-name>
    <servlet-class>com.woniu.servlet.JstlServlet</servlet-class>
</servlet>
<servlet-mapping>
    <servlet-name>jstl</servlet-name>
    <url-pattern>/jstl</url-pattern>
</servlet-mapping>
```

④ 浏览器访问 JSTL 请求，if 标签效果如图 1-24 所示。

图 1-24　if 标签效果

（2）set 标签，用于添加数据到作用域对象中。

```
<!-- set标签演示 -->
    <!-- 在page作用域中添加名称为a、值为hello的数据 -->
    <br><c:set var="a" value="hello"></c:set>
    ${pageScope.a }<br>
    <!-- 在session作用域中存储名称为b、值为hello的数据。var用于指定存储的名称；value用于指定存储
的值；scope用于指定存储的作用域，若不写，则默认为page作用域
    -->
    <c:set var="b" value="hello" scope="session"></c:set>
    ${sessionScope.b }<br>
```

访问 jstl 路径，set 标签效果如图 1-25 所示。

图 1-25　set 标签效果

（3）remove 标签，用于删除作用域中的数据。

在 jstl.jsp 中添加如下代码。

```
<!-- 删除所有作用域中name为a的数据 -->
<c:remove var="a"/>
<!-- 删除指定作用域中name为a的数据 -->
<c:remove var="b" scope="session"/>
${a }<br>
${b }<br></body>
```

访问 jstl 路径，remove 标签效果如图 1-26 所示。

图 1-26　remove 标签效果

在使用了 remove 标签之后，再次通过 EL 表达式获取作用域中的数据，页面中并没有显示出数据，由此可以证明数据已经从作用域中移除。

（4）choose 标签，类似于 if else 分支选择结构。

在 jstl.jsp 中添加如下代码。

```
<!-- choose标签 -->
    <c:choose>
        <c:when test="${age>25 }">太老</c:when>
        <c:when test="${age==5 }">刚刚好</c:when>
        <c:otherwise>太小</c:otherwise>
    </c:choose>
    <br>
```

访问 jstl 路径，choose 标签效果如图 1-27 所示。

以上就是 JSP 的常用技术，作为 Java Web 中发展非常成熟的技术，其在开发中占有举足轻重的地位，希望读者通过大量练习来掌握这些技术的使用方法。

图 1-27　choose 标签效果

1.1.3　过滤器

通过所学技术，读者已经可以进行 Web 程序的开发，但是有些业务是 Servlet 需要处理的，如登录验证、权限管理等，这些业务通常是每个 Servlet 都会判断的。

这样的业务，如果每一个 Servlet 都写一次，会造成代码的大量重复，这时可以使用过滤器（Filter）来完成。

Filter 是 Servlet 2.3 新增的技术，在开发 Web 应用时，可以编写一个 Java 类实现 Filter 接口，这个类就是一个过滤器，过滤器可以在请求到达 Servlet 之前对其进行预处理，也可以在请求离开时再次对 response 进行处理。Filter 的处理流程如图 1-28 所示。

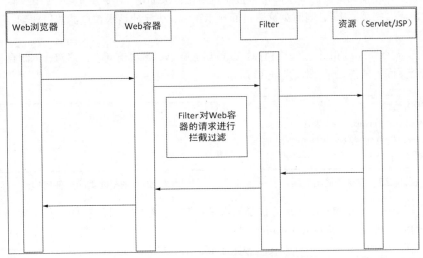

图 1-28　Filter 的处理流程

下面通过代码对 Filter 进行详细讲解。

（1）编写 Java 类，该类需要实现 Filter 接口，重写 doFilter 方法。

```
package com.woniu.filter;
import java.io.IOException;
import javax.servlet.*;
public class FirstFilter implements Filter{
    //此方法为初始化方法
    public void init(FilterConfig filterConfig) throws ServletException {
    }
    //重写doFilter方法，此方法为过滤方法
```

```java
public void doFilter(ServletRequest request, ServletResponse response, FilterChain chain)
        throws IOException, ServletException {
    System.out.println("拦截请求");
    /*FilterChain是过滤器链对象，一个请求会有一个完整的过滤器链，根据web.xml配置的顺序，将过
滤器组装为过滤器链执行doFilter方法，传入request、response对象，当前过滤器执行完成后，交由下一个过
滤器链上的过滤器继续过滤，否则请求不会向下执行 */
    chain.doFilter(request, response);
}
//销毁方法，同初始化方法
public void destroy() {
}
}
```

（2）编写web.xml，添加如下代码。

```xml
<!-- filter配置，配置结构和servlet相同 -->
<filter>
    <filter-name>first</filter-name>
    <filter-class>com.woniu.filter.FirstFilter</filter-class>
</filter>
<filter-mapping>
    <filter-name>first</filter-name>
    <!-- 这里配置为过滤所有的请求，* 表示通配符-->
    <url-pattern>/*</url-pattern>
</filter-mapping>
```

（3）任意访问一个前面配置的请求即可，控制台输出效果如图1-29所示。

控制台输出了"拦截请求"，同时浏览器页面显示正常，表示此请求被过滤器过滤到，并且整个请求处理链完整。

（4）将过滤器chain.doFilter方法注释掉，并重启服务器，再次访问jstl请求，控制台输出过滤器请求，但是浏览器页面无任何显示，如图1-30所示。

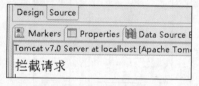
图1-29 控制台输出效果

图1-30 浏览器页面无任何显示

由此可知，FilterChain 的 doFilter 方法是请求是否继续执行的关键所在，在实际Web项目开发中，通常根据具体的业务决定是否执行 doFilter 方法。

（5）过滤器后置过滤，修改FirstFilter类中的doFilter方法，代码如下。

```java
public void doFilter(ServletRequest request, ServletResponse response, FilterChain chain)
        throws IOException, ServletException {
    System.out.println("First拦截请求前置部分");
    chain.doFilter(request, response);
    /* 在dofilter方法之后，可以再次对request和response对象进行处理，此时请求已经经过了Servlet
的处理，取出在JstlServlet类中绑定的name属性做此验证 */
    System.out.println(request.getAttribute("name"));;
    System.out.println("First拦截请求后置部分");
}
```

在 JstlServlet 的 service 方法中添加输出语句，浏览器再次访问 jstl 请求，过滤器前后过滤请求，如图1-31所示。

```
Tomcat v7.0 Server at localhost [Apache Tomcat] C:\Progra
First拦截请求前置部分
JstlServlet service方法运行
李四
First拦截请求后置部分
```

图 1-31　过滤器前后过滤请求

chain.doFilter 方法前的代码是在 Servlet 之前执行的，chain.doFilter 方法后的代码是在 Servlet 之后执行的。

（6）如果有多个过滤器过滤同一个请求，则过滤顺序是“先进后出”，即前置代码先执行的，后置代码会后执行。例如，有两个过滤器 A 和 B，A 前置代码先执行，B 前置代码后执行，则 B 的后置代码先执行，A 的后置代码后执行。创建类 SecondFilter，其 doFilter 方法代码如下。

```
public void doFilter(
          ServletRequest request,
          ServletResponse response, FilterChain chain)
          throws IOException, ServletException {
      System.out.println("Second拦截请求前置部分");
      chain.doFilter(request, response);
      System.out.println("Second拦截请求后置部分");
   }
```

在 web.xml 中添加如下配置代码。

```
<filter>
  <filter-name>second</filter-name>
  <filter-class>com.woniu.filter.SecondFilter</filter-class>
</filter>
<filter-mapping>
  <filter-name>second</filter-name>
  <url-pattern>/*</url-pattern>
</filter-mapping>
```

浏览器访问 jstl 请求，显示正常，Eclipse 控制台输出内容如图 1-32 所示。

```
First拦截请求前置部分
Second拦截请求前置部分
JstlServlet service方法运行
Second拦截请求后置部分
李四
First拦截请求后置部分
```

图 1-32　Eclipse 控制台输出内容

过滤器在实际项目开发中可以完成哪些工作呢？下面用登录验证案例为读者演示，在做案例之前，需要再次为读者详细介绍 Session。

众所周知，Web 程序前后端交互发送请求的协议是 HTTP，而 HTTP 的特点是短连接、无状态，但通常程序都是需要状态的，如登录的状态，用户是否登录表示操作是否可以完成，这对于一个程序而言非常重要。HTTP 一次请求，一次连接，响应之后就会断开连接，在这样的情况下，就需要使用 Session 保存用户的状态，其在一次会话中有效。

在 1.1 节演示的登录案例中，浏览器访问登录页面，输入用户名和密码，单击“登录”按钮，访问 login 请求，如果用户名和密码正确，则跳转到 index.jsp，如果不正确，则跳转到 login.jsp 页面，即 index.jsp 是一个需要登录之后才能访问的页面。但此时直接访问 index.jsp，发现是可以正常访问的，按照正常的业务逻辑，用户直接访问 index.jsp 时，应当加上一个判断，如果用户没有登录，则跳转到登录页面提示用户登录。在此案例的基础之上添加 LoginFilter 过滤所有请求，如果是 login 请求或 login.jsp 请求，则希望 LoginFilter 过滤器不过滤此请求。LoginFilter 的代码如下。

```
public void doFilter(ServletRequest request, ServletResponse response, FilterChain
chain)throws IOException, ServletException {
    //将ServletRequest转换为HttpServletRequest
    HttpServletRequest req = (HttpServletRequest)request;
    //将ServletResponse转换为HttpServletResponse
    HttpServletResponse resp = (HttpServletResponse)response;
    //获取到请求地址URI
    String uri = req.getRequestURI();
    //将第一个 '/' 截取掉
    uri = uri.substring(1);
    //去除路径的项目名称
    uri = uri.substring(uri.indexOf('/')+1);
    //如果请求为login或login.jsp，则执行不处理
    if("login".equals(uri) || "login.jsp".equals(uri)) {
        chain.doFilter(req, resp);
        return;
    }
    //获取到session对象
    HttpSession session = req.getSession();
    /*获取到绑定在session对象上的uname属性(此数据是在LoginServlet用户登录成功之后存储在session
对象中的数据)*/
    String uname = (String)session.getAttribute("uname");
    //若uname不为空或空字符串，则表示用户已经登录
    if(uname!=null && !"".equals(uname)) {
        chain.doFilter(req, resp);
        return;
    }
    /*如果用户没有登录并且请求路径不是login或login.jsp，则重定向到login.jsp页面*/
    resp.sendRedirect("login.jsp");
}
```

在 web.xml 中添加如下配置代码。

```
<filter>
    <filter-name>login</filter-name>
    <filter-class>com.woniu.filter.LoginFilter</filter-class>
</filter>
<filter-mapping>
    <filter-name>login</filter-name>
    <url-pattern>/*</url-pattern>
</filter-mapping>
```

直接访问 index.jsp 页面，发现跳转到了 login.jsp 页面，输入正确的用户名和密码，跳转到 index.jsp 页面。此时重新打开一个新的页面，访问 index.jsp，可以访问成功，说明过滤器对用户是否登录已经做了过滤。下面将继续对功能进行完善，添加退出登录功能。

修改 index.jsp，代码如下。

```
<body>
    Welcome ,${uname}! <a href="logout">退出登录</a>
</body>
```

新建 LogoutServlet，代码如下。

```
protected void service(HttpServletRequest req, HttpServletResponse resp) throws
ServletException, IOException {
    //退出登录功能
    //需要将session的信息清除，并重定向到登录页面
```

```
        HttpSession session = req.getSession();
        //使session失效
        session.invalidate();
        //重定向到登录页面
        resp.sendRedirect("login.jsp");
    }
```

此时正常登录，再次访问 index.jsp，访问成功；单击"退出"按钮，页面跳转到 login.jsp，再次直接访问 index.jsp，同样跳转到了 login.jsp 页面，即 session 中的数据已经被清除。

1.2 Servlet 项目实战

V1-6 Filter

本节使用 Servlet、JSP、Filter 完成一个权限控制案例，该案例中包含 MySQL 数据库、JDBC、数据库连接池等技术，主要目的是让读者体会到 Java Web 开发的实际运用。项目主要流程如下：由用户进行登录，登录成功之后进入主页面，主页面显示商品信息列表，并根据用户的权限显示用户对商品能够进行的操作，如删除、修改等，并提供退出登录功能。

1.2.1 开发环境搭建

创建 Maven 项目 lflsys，因为项目包含前后端交互，因此一定要选择打包方式为 war。同时，需要在 pom.xml 中导入所需的 JAR 包，具体代码如下。

```xml
<!-- Servlet -->
<dependency>
    <groupId>javax.servlet</groupId>
    <artifactId>servlet-api</artifactId>
    <version>2.5</version>
    <scope>provided</scope>
</dependency>
<!-- jstl -->
<dependency>
    <groupId>javax.servlet</groupId>
    <artifactId>jstl</artifactId>
    <version>1.2</version>
</dependency>
<!-- mysql-connector-java:JDBC -->
<dependency>
    <groupId>mysql</groupId>
    <artifactId>mysql-connector-java</artifactId>
    <version>5.1.6</version>
</dependency>
<!-- druid:数据库连接池 -->
<dependency>
    <groupId>com.alibaba</groupId>
    <artifactId>druid</artifactId>
    <version>1.0.12</version>
</dependency>
```

导入 JAR 包完成后，可以通过 Eclipse 将项目添加到 Tomcat 中，至此开发环境搭建完毕。

1.2.2 MySQL 数据库搭建

这里需要确保计算机安装了 MySQL 数据库。安装 MySQL 非常简单，网上教程也非常多，读者可以自行完成，安装完成之后即可进行数据库的设计。数据库详细设计如图 1-33 所示。

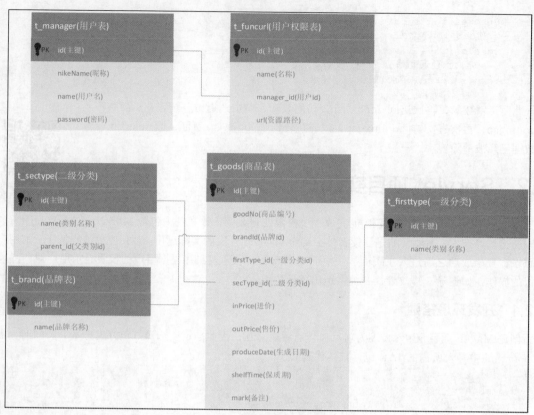

图 1-33　数据库详细设计

数据库中共设计了 6 张表。其中，用户管理分为两张表，一张为用户表，另一张为用户权限表，用户权限表对应用户可以进行的操作，分为 update、delete、add 3 个级别。其余 4 张表为商品表，分别是一级分类、二级分类、品牌表、商品表。本案例使用的数据均为测试数据。

数据库设计完成之后就可以开始后端和前端代码的编写工作，同时需要使用 JDBC 和数据库连接池连接到数据库。为了使代码具有可维护性，需要将连接数据库的必要数据写入一个 druid.properties 文件，代码如下。

```
driveName=com.mysql.jdbc.Driver
url=jdbc:mysql://127.0.0.1:3306/lflsys
user=root
password=338D281AE647724B0483DCCB73DA2D51
initialSize=10
maxActive=50
minIdle=15
maxWait=3000
```

创建 com.lfl.util 包，在包中创建 DBUtil 类，代码如下。

```
package com.lfl.util;

import java.io.InputStream;
import java.sql.Connection;
import java.sql.SQLException;
import java.util.Properties;

import com.alibaba.druid.pool.DruidDataSource;
```

```java
public class DBUtil {
    private static String driveNamen;
    private static String url;
    private static String user;
    private static String password;

    /**连接池对象*/
    private static DruidDataSource druid;
    private static Integer initialSize;
    private static Integer maxActive;
    private static Integer minIdle;
    private static Integer maxWait;

    static{
        try {
            Properties prop = new Properties();
            InputStream in = DBUtil.class.getClassLoader().
                    getResourceAsStream("com/lfl/util/druid.properties");
            prop.load(in);
            druid = new DruidDataSource();
            //获取基本信息
            driveNamen = prop.getProperty("driveName");
            url = prop.getProperty("url");
            user = prop.getProperty("user");
            password = prop.getProperty("password");
            //获取连接池信息
            initialSize =
                    Integer.parseInt(prop.getProperty("initialSize"));
            maxActive = Integer.parseInt(prop.getProperty("maxActive"));
            minIdle = Integer.parseInt(prop.getProperty("minIdle"));
            maxWait = Integer.parseInt(prop.getProperty("maxWait"));

            //配置基本信息
            druid.setDriverClassName(driveNamen);
            druid.setUrl(url);
            druid.setUsername(user);
            druid.setPassword(Security.AES.decrypt(password, "woniu"));
            //配置连接池信息
            druid.setInitialSize(initialSize);
            druid.setMaxActive(maxActive);
            druid.setMinIdle(minIdle);
            druid.setMaxWait(maxWait);
            in.close();
        } catch (Exception e) {
            e.printStackTrace();
        }
    }
    /**
     * 获取数据库连接
     * @return
     */
    public static Connection getConnection(){
        Connection conn = null;
        try {
            conn = druid.getConnection();
```

```
        } catch (SQLException e) {
            System.out.println("获取连接失败....");
            e.printStackTrace();
        }
        return conn;
    }

    /**
     * 关闭数据库连接
     * @param conn
     */
    public static void close( Connection conn ){
        try {
            if( conn != null )conn.close();
        } catch (SQLException e) {
            e.printStackTrace();
        }
    }
    //测试
    public static void main(String[] args) {
        System.out.println(DBUtil.getConnection());
    }
}
```

在数据库连接中使用到了加密和解密类 Security，代码如下。

```
package com.lfl.util;

import java.io.UnsupportedEncodingException;
import java.security.InvalidKeyException;
import java.security.NoSuchAlgorithmException;
import java.security.SecureRandom;

import javax.crypto.*;
import javax.crypto.spec.SecretKeySpec;

public class Security {
    //AES对称加密算法
    public static class AES {
        /**
         * AES加密
         * @param content 为需要加密的内容
         * @param password 为加密密码
         * @return 为十六进制字符串
         */
        public static String encrypt(String content, String password) {
            try {
                KeyGenerator kgen = KeyGenerator.getInstance("AES");
                kgen.init(128, new SecureRandom(password.getBytes()));
                SecretKey secretKey = kgen.generateKey();
                byte[] enCodeFormat = secretKey.getEncoded();
                SecretKeySpec key = new SecretKeySpec(enCodeFormat, "AES");
                Cipher cipher = Cipher.getInstance("AES");//创建密码器
                byte[] byteContent = content.getBytes("utf-8");
                cipher.init(Cipher.ENCRYPT_MODE, key);//初始化
                byte[] result = cipher.doFinal(byteContent);
                return parseByte2HexStr(result);//加密
```

```java
        } catch (Exception e) {
            e.printStackTrace();
        }
        return null;
    }
    //解密数据，通常是对配置文件中的加密密码进行解密
    public static String decrypt(String ciphertext, String password) {
        byte[] content = parseHexStr2Byte(ciphertext);
        try {
            KeyGenerator kgen = KeyGenerator.getInstance("AES");
            kgen.init(128, new SecureRandom(password.getBytes()));
            SecretKey secretKey = kgen.generateKey();
            byte[] enCodeFormat = secretKey.getEncoded();
            SecretKeySpec key = new SecretKeySpec(enCodeFormat, "AES");
            Cipher cipher = Cipher.getInstance("AES");//创建密码器
            cipher.init(Cipher.DECRYPT_MODE, key);//初始化
            byte[] result = cipher.doFinal(content);//解密
            return new String(result); //转换为字符串并返回
        } catch (Exception e) {
            e.printStackTrace();
        }
        return null;
    }
    // 将二进制转换为十六进制
    private static String parseByte2HexStr(byte buf[]) {
        StringBuffer sb = new StringBuffer();
        for (int i = 0; i < buf.length; i++) {
            String hex = Integer.toHexString(buf[i] & 0xFF);
            if (hex.length() == 1) {
                hex = '0' + hex;
            }
            sb.append(hex.toUpperCase());
        }
        return sb.toString();
    }
    //将十六进制转换为二进制
    private static byte[] parseHexStr2Byte(String hexStr) {
        if (hexStr.length() < 1)
            return null;
        byte[] result = new byte[hexStr.length()/2];
        for (int i = 0;i< hexStr.length()/2; i++) {
            int high = Integer.parseInt(
                    hexStr.substring(i*2, i*2+1), 16);
            int low = Integer.parseInt(
                    hexStr.substring(i*2+1, i*2+2), 16);
            result[i] = (byte) (high * 16 + low);
        }
        return result;
    }
}
//测试
public static void main(String[] args) {
    //root:原文密码，加密生成的密文手动存入数据库配置文件
    //woniu:密钥
    System.out.println(Security.AES.encrypt("root","woniu"));
```

```
    //338D281AE647724B0483DCCB73DA2D51
    //6CC47F2BAB46E79EB50236FAE973ECBA
    System.out.println(
            Security.AES.decrypt("K4dA5NpSEzStvxTMiZF/QQ", "woniu"));
    }
}
```

测试 DBUtil 中获取连接的方法，正常输出即表示代码运行成功，可以进入下一步编码工作。

1.2.3　Servlet 请求处理

创建 com.lfl.action 包，用于存放所有的 Servlet 类，包括处理登录逻辑的 LoginAction，代码如下。

```
package com.lfl.action;

import java.io.IOException;
import java.util.HashMap;
import java.util.List;
import java.util.Map;

import javax.servlet.ServletException;
import javax.servlet.http.*;

import com.lfl.dao.FuncDao;
import com.lfl.dao.ManagerDao;
import com.lfl.dao.impl.FuncDaoImpl;
import com.lfl.dao.impl.ManagerDaoImpl;
import com.lfl.model.Func;
import com.lfl.model.Manager;
import com.tools.encryption.MD5;

//登录数据处理
public class LoginAction extends HttpServlet{
    @Override
    protected void service(HttpServletRequest req, HttpServletResponse resp) throws
ServletException, IOException {
        //获取页面数据
        String name = req.getParameter("name");
        String password = req.getParameter("password");
        try {
            HttpSession session = req.getSession(false);
            //以下是数据库操作
            ManagerDao md = new ManagerDaoImpl();
            Manager manager = md.queryManager(
                        name, MD5.convert32(password));
            if( manager != null ){//合法管理员
                //将登录用户绑定到session中
                session.setAttribute("name", manager);
                System.out.println(session.getId());
                //查询出用户具有的权限，并存入session作用域
                FuncDao fun = new FuncDaoImpl();
                List<Func> funList = fun.queryFuncUrl(manager.getId());
                session.setAttribute("funList", funList);
                req.getRequestDispatcher(
                        "indexPage.action").forward(req, resp);
                return;
            }else{
```

```
                req.getSession().setAttribute(
                                "loginInfo", "用户名或密码错误");
                resp.sendRedirect("login.jsp");
            }
        } catch (Exception e) {
            System.out.println("xxxxx");
            e.printStackTrace();
        }
    }
}
```

登录成功之后转发到 indexPage.action 请求，此请求由 IndexPageAction 处理，代码如下。

```
package com.lfl.action;

import java.io.IOException;
import java.util.List;

import javax.servlet.ServletException;
import javax.servlet.http.*;

import com.lfl.dao.FuncDao;
import com.lfl.dao.GoodsDao;
import com.lfl.dao.impl.FuncDaoImpl;
import com.lfl.dao.impl.GoodsDaoImpl;
import com.lfl.model.Goods;

public class IndexPageAction extends HttpServlet{
    @Override
    protected void service(HttpServletRequest req, HttpServletResponse resp) throws ServletException, IOException {
        try {
            //查询所有库存产品
            GoodsDao gd = new GoodsDaoImpl();
            List<Goods> list = gd.queryAll();
            req.getSession(false).setAttribute("goodsList", list);
            //转发到主页
            resp.sendRedirect("page/index.jsp");
        } catch (Exception e) {
            e.printStackTrace();
        }
    }
}
```

前端页面 login.jsp 的代码如下。

```
<%@ page language="java" import="java.util.*" pageEncoding="UTF-8"
contentType="text/html;charset=UTF-8"%>
<%
//获取项目的基础路径，使用绝对路径时通常加上basePath作为开头
String path = request.getContextPath();
String basePath = request.getScheme()+"://"
    +request.getServerName()+":"+request.getServerPort()+path+"/";
%>
<!DOCTYPE html>
<html>
    <head>
        <title>登录</title>
        <meta name="content-type" content="text/html; charset=UTF-8">
```

```html
            <meta charset="UTF-8"/>
</head>
<style>
    body{
        background-image: url("image/1fl.jpg");
        background-size: 100%,100%;
    }
    #image{
        width: 565px;
        height: 265px;
        border-radius: 10px;
        margin: 220px auto 0;
        border: solid #0098c7 1px;
        background-image: url("image/loginbg.gif");
    }
    #name,#input,#user{
        width: 160px;
        height: 140px;
        float: left;
        margin: 100px 0 0 0;
    }
    #name{
        margin-left: 20px;
    }
    .name{
        width: 90px;
        height: 40px;
        text-align: center;
        line-height: 40px;
        margin-left: 70px;
    }
    .input{
        width: 150px;
        height: 40px;
        text-align: center;
        line-height: 40px;
    }

    #text{
        background-image: url("image/usernamebg.gif");
    }
    #password{
        background-image: url("image/passwordbg.gif");
    }
    #text,#password{
        border: 0;
        background-size: 100% 100%;
        padding-left: 20px;
        width: 160px;
        height: 25px;
        border-radius: 5px;
    }
    #user{
        margin: 100px 0 0  60px;
    }
```

```
            #below{
                width: 565px;
                height: 20px;
                margin-top: 240px;
                text-align: center;
                font-size: 12px;
            }
            .subBtn:HOVER {
                cursor:pointer;
            }
            #codeImg:HOVER {
                cursor:pointer;
            }
        </style>
        <script type="text/javascript">
            function subForm(){
                var formTag = document.getElementById("form");
                formTag.submit();
            }
            function changeCode(imgTag){
                imgTag.src = imgTag.src + "?code=" + Math.random();
            }
        </script>
    <body>
        <form id="form" action="login.action" method="post">
            <div id="image">
                <div id="name">
                    <div class="name">用户名：</div>
                    <div class="name">密    码：</div>
                    <div class="name" style="color:red">${ loginInfo }</div>
                </div>
                <div id="input">
                    <div class="input"><input type="text" id="text"
                             name="name" value="${ name }"/></div>
                    <div class="input"><input type="password" id="password"
                            name="password" value="${ password }"/></div>
                </div>
                <div id="user">
                    <div class="subBtn" onclick="subForm();">
                        <img src="image/loginbt.gif" />
                    </div>
                </div>
            </div>
        </form>
    </body>
</html>
```

登录成功之后最终跳转到 index.jsp 页面，代码如下。

```
<%@ page language="java" import="java.util.*,com.lfl.model.*"
pageEncoding="UTF-8" contentType="text/html;charset=UTF-8"%>
<%
String path = request.getContextPath();
String basePath =
      request.getScheme()+"://"+request.getServerName()+":"+
      request.getServerPort()+path+"/";
request.setAttribute("basePath", basePath);
```

```jsp
%>
<%@taglib uri="http://java.sun.com/jsp/jstl/core" prefix="c" %>
<!DOCTYPE html>
<html>
    <head>
        <title>笠芙莱管理系统</title>

        <meta name="keywords" content="keyword1,keyword2,keyword3">
        <meta name="description" content="this is my page">
        <meta name="content-type" content="text/html; charset=UTF-8">
        <meta charset="UTF-8"/>
    </head>
    <style>
        body,html{
        margin: 0;
        width: 100%;
        height: 100%;
        }
        #top{/*设置页面页眉的样式*/
            width: 100%;
            height: 8%;
            background-image: url("../image/headerbg.gif");
            background-size: 100% 100%;
            min-height: 80px;
            min-width: 1000px;
        }
        #left{ /*设置左边栏的样式*/
            width: 14%;
            height: 100%;
            background-image: url("../image/leftbg.gif");
            background-size: 100% 100%;
            float: left;
        }
        #left-top{
            width: 90%;
            height: 50%;
            margin-top: 20px;
        }
        .left{
            width:100%;
            height: 8%;
            color: #09c;
            font-size: 16px;
            text-align: center;
            margin: 0 auto 0;
        }
        #right{/*设置右边栏的样式*/
            width: 85.9%;
            height:100%;
            margin-left: 0.1%;
            float: left;
        }
    </style>
    <script type="text/javascript" src="../js/jquery-1.7.2.js"></script>
    <script type="text/javascript">
```

```html
        function deleteData(id){
            if(confirm("确定删除？")){
                location.href="deleteGoods.action?id="+id;
            }
        }
        window.onbeforeunload = function(){
            $.ajax({
                url:"logout.action"
            });
        }
    </script>
    <body>
        <div id="top" style="padding-left: 10px;">
            欢迎:${ name.nikeName }
                
            <a href="logout.action">退出系统</a>
        </div>
        <div id="right" style="text-align: center;">
            <h1>库存信息</h1>
            <table border="1" width="80%" align="center">
                <tr>
                    <td>编号</td>
                    <td>产品名称</td>
                    <td>一级分类</td>
                    <td>二级分类</td>
                    <td>进价</td>
                    <td>售价</td>
                    <td>保质期</td>
                    <td>操作</td>
                </tr>
                <!-- 取出商品列表 -->
                <c:forEach items="${goodsList }" var="goods" varStatus="s">
                    <tr>
                        <td>${s.count }</td>
                        <td>${goods.name }</td>
                        <td>${goods.firstType }</td>
                        <td>${goods.secType }</td>
                        <td>${goods.inPrice }</td>
                        <td>${goods.outPrice }</td>
                        <td>${goods.shelfTime }</td>
                        <td>
                            <!-- 取出用户的权限,如果用户具有相应的权限,则显示操作按钮 -->
                            <c:forEach var="fun" items="${funList }">
                                <c:if test="${fun.url=='delete' }">
                                    <a href="${basePath }deleteGoods.action?id=${goods.id }">删除产品</a>
                                </c:if>
                                   
                                <c:if test="${fun.url=='update' }">
                                    <a>修改产品</a>
                                </c:if>
                            </c:forEach>
                        </td>
                    </tr>
                </c:forEach>
```

```
            </table>
        </div>
    </body>
</html>
```

JDBC 是比较底层的技术，在实际开发工作中，不会直接使用 JDBC 进行开发，而是会对 JDBC 进行一层或多层的封装，主要目的是将数据库的数据转换为 Java 中的对象。例如，对于数据库 Manager 表，使用 JDBC 进行查询，获取到的是一堆数据。封装的作用就是将这一堆数据转换为 Java 对象。

创建接口 ManagerDao，代码如下。

```java
package com.lfl.dao;

import com.lfl.model.Manager;

public interface ManagerDao {
    //根据用户名和密码查询Manager对象
    public Manager queryManager(String name, String password)
                                                throws Exception;
}
```

创建 ManagerDao 接口的实现类 ManagerDaoImpl，代码如下。

```java
package com.lfl.dao.impl;

import java.sql.Connection;
import java.sql.PreparedStatement;
import java.sql.ResultSet;
import java.sql.SQLException;

import com.lfl.dao.ManagerDao;
import com.lfl.model.Manager;
import com.lfl.util.DBUtil;

public class ManagerDaoImpl implements ManagerDao{

    @Override
    public Manager queryManager(String name, String password)
            throws SQLException {
        Manager manager = null;
        Connection conn = null;
        try {
            conn = DBUtil.getConnection();
            System.out.println();
            String sql =
                " select * from t_manager where name=? and password=? ";
            PreparedStatement ps = conn.prepareStatement(sql);
            ps.setString(1, name);
            ps.setString(2, password);
            ResultSet rs = ps.executeQuery();
            if( rs.next() ){ //用户合法
                manager = new Manager();
                manager.setId(rs.getInt("id"));
                manager.setNikeName(rs.getString("nikeName"));
                manager.setName(rs.getString("name"));
                manager.setPassword(rs.getString("password"));
                manager.setMgrLevel(rs.getInt("mgrLevel"));
            }
```

```
            } catch (SQLException e) {
                //处理异常
                throw e;
            }finally{
                DBUtil.close(conn);
            }
            return manager;
    }
}
```

商品查询 GoodsDao 接口的代码如下。

```
package com.lfl.dao;

import java.util.List;

import com.lfl.model.Goods;

public interface GoodsDao {
    //查询所有库存产品
    public List<Goods> queryAll() throws Exception;
    //删除产品
    public void delete(Integer id) throws Exception;
    //根据id查询一个产品
    public Goods findById(Integer id) throws Exception;
}
```

GoodsDao 接口的实现类 GoodsDaoImpl 的代码如下。

```
package com.lfl.dao.impl;

import java.sql.Connection;
import java.sql.PreparedStatement;
import java.sql.ResultSet;
import java.sql.SQLException;
import java.util.ArrayList;
import java.util.List;

import com.lfl.dao.GoodsDao;
import com.lfl.model.Goods;
import com.lfl.model.Manager;
import com.lfl.util.DBUtil;

public class GoodsDaoImpl implements GoodsDao{

    @Override
    public List<Goods> queryAll() throws Exception {
        List<Goods> list = new ArrayList<Goods>();
        Goods goods = null;
        Connection conn = null;
        try {
            conn = DBUtil.getConnection();
            String sql = " select gd.id,gd.goodNo, br.name bName, "
                    + "fst.name fName, sec.name sName, gd.inPrice, gd.outPrice"
                    + ",gd.shelfTime "+ " from 't_goods' gd "
                    + " left join 't_brand' br on gd.brand_id = br.id "
                    + " left join 't_firsttype' fst on gd.firstType_id "
                    + "= fst.id left join 't_sectype' sec on "
                    + "sec.id = gd.secType_id ";
```

```java
            PreparedStatement ps = conn.prepareStatement(sql);
            ResultSet rs = ps.executeQuery();
            while( rs.next() ){ //用户合法
                goods = new Goods();
                goods.setId(rs.getInt("id"));
                goods.setGoodNo(rs.getString("goodNo"));
                goods.setName(rs.getString("bName"));
                goods.setFirstType(rs.getString("fName"));
                goods.setSecType(rs.getString("sName"));
                goods.setInPrice(rs.getInt("inPrice"));
                goods.setOutPrice(rs.getInt("outPrice"));
                goods.setShelfTime(rs.getInt("shelfTime"));
                list.add(goods);
            }
    } catch (SQLException e) {
        //处理异常
        throw e;
    }finally{
        DBUtil.close(conn);
    }
    return list;
}
@Override
public void delete(Integer id) throws Exception {
    Connection conn = null;
    try {
        conn = DBUtil.getConnection();
        String sql = " delete from 't_goods' where id = ? ";
        PreparedStatement ps = conn.prepareStatement(sql);
        ps.setInt(1, id);
        ps.execute();
    } catch (SQLException e) {
        //处理异常
        throw e;
    }finally{
        DBUtil.close(conn);
    }
}

@Override
public Goods findById(Integer id) throws Exception {
    Connection conn = null;
    Goods goods = null;
    try {
        conn = DBUtil.getConnection();
        String sql = " select * from t_goods where id= ? ";
        PreparedStatement ps = conn.prepareStatement(sql);
        ps.setInt(1, id);
        ResultSet rs = ps.executeQuery();
        //处理结果集
        if( rs.next() ){
            goods = new Goods();
            goods.setId(rs.getInt("id"));
            goods.setGoodNo(rs.getString("goodNo"));
            goods.setBrand_id(rs.getInt("brand_id"));
```

```
                goods.setSecType_id(rs.getInt("secType_id"));
                goods.setInPrice(rs.getInt("inPrice"));
            }
        } catch (SQLException e) {
            //处理异常
            throw e;
        }finally{
            DBUtil.close(conn);
        }
        return goods;
    }
}
```

权限查询 FuncDao 接口的代码如下。

```
package com.lfl.dao;

import java.util.List;

import com.lfl.model.Func;

public interface FuncDao {
    //根据用户ID查询出该用户具有的所有权限
    public List<Func> queryFuncUrl( Integer managerID ) throws Exception;
}
```

FuncDao 接口的实现类 FuncDaoImpl 的代码如下。

```
package com.lfl.dao.impl;

import java.sql.Connection;
import java.sql.PreparedStatement;
import java.sql.ResultSet;
import java.sql.SQLException;
import java.util.ArrayList;
import java.util.List;

import com.lfl.dao.FuncDao;
import com.lfl.model.Func;
import com.lfl.util.DBUtil;

public class FuncDaoImpl implements FuncDao{

    @Override
    public List<Func> queryFuncUrl( Integer managerID ) throws Exception {
        List<Func> list = new ArrayList<Func>();
        Func func = null;
        Connection conn = null;
        try {
            conn = DBUtil.getConnection();
            //查询出用户的所有权限
            String sql = " select id ,name, url, manager_id"
                       + " from t_funcurl"
                       + " where manager_id = ? ";
            PreparedStatement ps = conn.prepareStatement(sql);
            ps.setInt(1, managerID);
            ResultSet rs = ps.executeQuery();
            while( rs.next() ){
                //权限对应的Func类
```

```java
                func = new Func();
                func.setId(rs.getInt("id"));
                func.setName(rs.getString("name"));
                func.setUrl(rs.getString("url"));
                func.setManager_id(rs.getInt("manager_id"));
                list.add(func);
            }
        } catch (SQLException e) {
            //处理异常
            throw e;
        }finally{
            DBUtil.close(conn);
        }
        return list;
    }
}
```

代码中出现的 Manager、Func、Goods、Brands 类为数据库表对应的 JavaBean 类，这里展示 Goods 类的代码，其他类的代码都是按照数据库对应的表字段编写的，读者可以自行完成。

```java
package com.lfl.model;

import java.io.Serializable;

public class Goods implements Serializable{
    private Integer id;
    private String goodNo;
    private String name;
    private String firstType;
    private String secType;
    private Integer inPrice;
    private Integer outPrice;
    private Integer shelfTime;
    private Integer brand_id;
    private Integer firstType_id;
    private Integer secType_id;

    //getters和setters

}
```

删除产品处理 Servlet 类 DeleteGoodsAction 的代码如下。

```java
package com.lfl.action;

import java.io.IOException;

import javax.servlet.ServletException;
import javax.servlet.http.HttpServlet;
import javax.servlet.http.HttpServletRequest;
import javax.servlet.http.HttpServletResponse;
import com.lfl.dao.GoodsDao;
import com.lfl.dao.impl.GoodsDaoImpl;

public class DeleteGoodsAction extends HttpServlet{
    @Override
    protected void service(HttpServletRequest req, HttpServletResponse resp) throws ServletException, IOException {
        try {
```

```
            String idStr = req.getParameter("id");
            if( idStr == null )throw new Exception("ID是空的！");
            Integer id = Integer.parseInt(idStr);
            
            GoodsDao gd = new GoodsDaoImpl();
            gd.delete(id);
            resp.sendRedirect("indexPage.action");
        } catch (Exception e) {
            e.printStackTrace();
        }
    }
}
```

退出登录 LogoutAction 的代码如下。

```
package com.lfl.action;

import java.io.IOException;

import javax.servlet.ServletException;
import javax.servlet.http.*;

public class LogoutAction extends HttpServlet{
    @Override
    protected void service(HttpServletRequest req, HttpServletResponse resp) throws ServletException, IOException {
        try {
            //从会话中销毁session
            req.getSession().invalidate();
            resp.sendRedirect("login.jsp");
        } catch (Exception e) {
            e.printStackTrace();
        }
    }
}
```

在 web.xml 中配置请求处理映射关系，代码如下。

```
<!-- 登录请求 -->
<servlet>
    <servlet-name>login</servlet-name>
    <servlet-class>com.lfl.action.LoginAction</servlet-class>
</servlet>
<servlet-mapping>
    <servlet-name>login</servlet-name>
    <url-pattern>/login.action</url-pattern>
</servlet-mapping>
<!-- 访问主页 -->
<servlet>
    <servlet-name>indexPage</servlet-name>
    <servlet-class>com.lfl.action.IndexPageAction</servlet-class>
</servlet>
<servlet-mapping>
    <servlet-name>indexPage</servlet-name>
    <url-pattern>/indexPage.action</url-pattern>
</servlet-mapping>
<!-- 删除产品 -->
<servlet>
    <servlet-name>deleteGoods</servlet-name>
```

```
        <servlet-class>com.lfl.action.DeleteGoodsAction</servlet-class>
    </servlet>
    <servlet-mapping>
        <servlet-name>deleteGoods</servlet-name>
        <url-pattern>/deleteGoods.action</url-pattern>
    </servlet-mapping>

    <!-- 退出系统 -->
    <servlet>
        <servlet-name>logout</servlet-name>
        <servlet-class>com.lfl.action.LogoutAction</servlet-class>
    </servlet>
    <servlet-mapping>
        <servlet-name>logout</servlet-name>
        <url-pattern>/logout.action</url-pattern>
    </servlet-mapping>
```

使用用户账号 zhangsan、密码 123456 登录，登录成功后，商品列表显示效果如图 1-34 所示。

库存信息

编号	产品名称	一级分类	二级分类	进价	售价	保质期	操作
1	百雀羚	女士	保湿	500	999	24	修改产品 删除产品
2	大宝	男士	控油	500	999	24	修改产品 删除产品

图 1-34　商品列表显示效果

zhangsan 具有修改和删除权限，所以在操作列表中可以单击进行操作（本案例仅实现了删除功能，修改功能留给读者实现），退出后使用用户账号 lisi、密码 123456 进行登录，而 lisi 只有修改权限，并没有删除权限，lisi 账号登录成功后，页面显示内容如图 1-35 所示。

库存信息

编号	产品名称	一级分类	二级分类	进价	售价	保质期	操作
1	百雀羚	女士	保湿	500	999	24	修改产品
2	大宝	男士	控油	500	999	24	修改产品

图 1-35　页面显示内容

删除功能是根据商品的 ID 进行删除操作，请求路径为 deleteGoods.Action，这里不使用前端单击操作而是直接发送请求，并传入 ID 数据，这样就可以绕过前端直接将数据删除，这需要在后端进行权限判断，这里使用了 Filter 进行判断。

1.2.4　Filter 权限控制

本案例提供两个过滤器：一个是前面案例中使用过的登录验证过滤器，另一个是用户权限验证过滤器。

登录验证过滤器 MyFilter 的代码如下。

```
package com.lfl.filter;

import java.io.IOException;
import java.util.ArrayList;
import java.util.List;
import javax.servlet.*;
```

```java
import javax.servlet.http.*;
import javax.servlet.http.HttpServletResponse;
import javax.servlet.http.HttpSession;

public class MyFilter implements Filter {
    //不需要登录验证的请求
    private List<String> actionName;
    @Override
    public void destroy() {
    }
    @Override
    public void doFilter(ServletRequest request, ServletResponse response, FilterChain chain)
            throws IOException, ServletException {
        HttpServletRequest req = (HttpServletRequest) request;
        HttpServletResponse resp = (HttpServletResponse) response;
        //通过过滤器设置编码
        req.setCharacterEncoding("UTF-8");
        resp.setCharacterEncoding("UTF-8");
        resp.setContentType("text/html;charset=UTF-8");
        /* 获取当前进入过滤器的请求，与集合中的元素对比，如果当前请求保存在集合中，则直接放过 */
        String uri = req.getRequestURI();
        String nowAction = uri.substring(uri.lastIndexOf("/") + 1);
        //不过滤的请求
        for (String actionStr : actionName) {
            if (nowAction.equals(actionStr)) {
                chain.doFilter(req, resp);
                return;
            }
        }
        //获取session中的登录用户,判断该用户是否经过登录的流程
        HttpSession session = req.getSession(false);
        System.out.println(session.getId());
        Object obj = session.getAttribute("name");
        System.out.println(nowAction);
        System.out.println(obj);
        if (obj == null) { //未经过登录流程
            System.out.println("该用户没有登录！");
            req.getRequestDispatcher("login.jsp").forward(req, resp);
            return;
        }
        //放过请求
        chain.doFilter(req, resp);
    }
    @Override
    public void init(FilterConfig filterConfig) throws ServletException {
        //初始化不需要登录验证的请求
        actionName = new ArrayList<String>();
        actionName.add("loginPage.action");
        actionName.add("login.action");
        actionName.add("logout.action");
    }
}
```

用户权限验证过滤器 JurFilter 的代码如下。

```java
package com.lfl.filter;
```

```java
import java.io.IOException;
import java.util.List;

import javax.servlet.*;
import javax.servlet.http.*;

import com.lfl.model.Func;

//用户权限验证过滤器
public class JurFilter implements Filter{
    @Override
    public void destroy() {
    }
    @Override
    public void doFilter(
         ServletRequest arg0, ServletResponse arg1, FilterChain arg2)
          throws IOException, ServletException {
        HttpServletRequest req = (HttpServletRequest) arg0;
        HttpServletResponse resp = (HttpServletResponse) arg1;
        System.out.println(
            "JurFilter:"+req.getSession().getAttribute("loginManager"));
        /*获取当前进入过滤器的请求,与集合中的元素进行对比,如果当前请求保存在集合中,则直接放过 */
        String uri = req.getRequestURI();
        String nowAction = uri.substring(uri.lastIndexOf("/") + 1);
        /*判断是否需要权限验证,如果请求前缀以update、delete、add开头,则判断用户是否具有相应的
操作权限 */
        if(nowAction.startsWith("add") ||
          nowAction.startsWith("delete") ||
          nowAction.startsWith("update")) {
             HttpSession session = req.getSession();
            //权限验证
            boolean isURLFunc = false;
            List<Func> funList = (List<Func>)
                    session.getAttribute("funList");
            System.out.println(funList);
            for (Func f : funList) {
                if (nowAction.startsWith(f.getUrl())) {
                    isURLFunc = true;
                }
            }
            //若没有权限,则跳转到错误页面
            System.out.println(isURLFunc?"有权限":"无权限");
            if (!isURLFunc) {
                req.getRequestDispatcher(
                    "WEB-INF/page/error.jsp").forward(req, resp);
                return;
            }
        }
        arg2.doFilter(req, resp);
    }
    @Override
    public void init(FilterConfig arg0) throws ServletException {
    }
}
```

此时，如果用户跳过前端，则直接发送删除请求；如果该用户并不具有删除权限，则会跳转到 error.jsp 页面，代码如下。

```jsp
<%@ page language="java" import="java.util.*" pageEncoding="UTF-8"%>
<%
String path = request.getContextPath();
String basePath = request.getScheme()+"://"
+request.getServerName()+":"+request.getServerPort()+path+"/";
%>
<!DOCTYPE HTML PUBLIC "-//W3C//DTD HTML 4.01 Transitional//EN">
<html>
  <head>
    <base href="<%=basePath%>">
    <title>My JSP 'error.jsp' starting page</title>
  </head>
  <body>
    不具有权限！
  </body>
</html>
```

本章主要介绍了 JavaEE 企业级开发的基础知识 Servlet，Servlet 是 Java Web 中必不可少的组件。尽管现在企业都会使用框架进行开发，但是框架都基于底层技术，后续介绍的 Spring MVC 框架就是基于 Servlet 的开发框架。

第2章

JavaEE框架开发——SSM

本章导读

■ 本章主要介绍企业级开发框架Spring、Spring MVC 和 MyBatis 的应用，使读者掌握企业应用中对框架的使用。

学习目标

（1）熟练运用MyBatis完成对数据的持久化。
（2）熟练运用Spring MVC处理Web请求。
（3）熟练运用Spring提供的IoC容器及AOP编程方式。

2.1 MyBatis 概述

2.1.1 了解 MyBatis

MyBatis 是一个优秀的持久层框架，它以面向 SQL 的形式支持定制化的 SQL、存储过程及高级映射。MyBatis 是对原有 JDBC 的包装，但避免了几乎所有的 JDBC 代码、手动设置参数及获取结果集，MyBatis 提供了基于 XML 或注解的方式来配置和映射原生信息，将接口和简单的 Java 对象（Plain Ordinary Java Objects，POJO）映射为数据库中的数据。MyBatis 架构如图 2-1 所示。

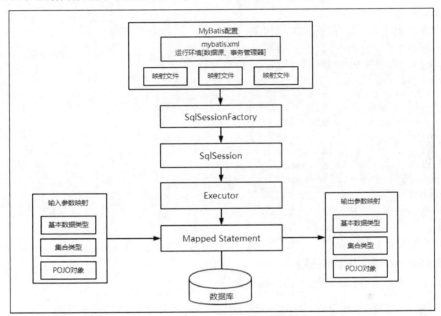

图 2-1　MyBatis 架构

其中，各要素的说明如下。

（1）mybatis.xml 配置文件：此文件为 MyBatis 的全局配置文件，配置了 MyBatis 的运行环境、映射文件、别名等信息。

（2）映射文件：映射文件配置了操作数据库的 SQL 语句，此文件需要在 MyBatis 的配置文件中进行配置加载。

（3）SqlSessionFactory：即会话工厂，该对象线程安全，主要用于创建 SqlSession 对象。

（4）SqlSession：即会话，是 MyBatis 提供给应用层的一个接口，所有的数据库操作都需要通过 SqlSession 进行，该接口有一个默认的实现类 DefaultSqlSession。

（5）Executor：执行器接口，该接口是 MyBatis 操作数据库底层的接口，Executor 接口有两个执行器，一个是基本执行器，另一个是缓存执行器。

（6）Mapped Statement：是 MyBatis 的一个底层封装对象，它包装了 MyBatis 配置信息及 SQL 映射信息等。映射文件中的每一个 SQL 都对应一个 Mapped Statement 对象，SQL 的 ID 即是 Mapped Statement 的 ID。

Mapped Statement 对 SQL 执行输入参数进行定义，包括集合类型、基本类型、POJO 对象，Executor 通过 Mapped Statement 在执行 SQL 前将输入的 Java 对象映射到 SQL 中，参数映射通过 JDBC 中的 PreparedStatement 完成。

Mapped Statement 对 SQL 执行输出结果进行定义，包括集合类型、基本类型、POJO 对象，Executor 通过 Mapped Statement 在执行 SQL 后将输出结果映射到 Java 对象中，输出结果映射过程相当于 JDBC 编程中结果集的解析处理。

2.1.2 MyBatis 数据持久化

通过前面的学习，读者可以了解到 MyBatis 实质上是对 JDBC 的包装，但省去了结果集的处理和参数的绑定过程，下面开始逐步通过 MyBatis 来完成对数据的持久化。本书的后续章节都将采用 Maven 来完成对项目的构建，如果读者对 Maven 不太熟悉，则可参考蜗牛学院官网相关介绍。

V2-1 MyBatis 简介

1．准备工作

创建数据库及用户表，打开 MySQL 数据库管理工具，并执行以下命令。

```sql
drop database if EXISTS mybatis;
create database mybatis;
use mybatis;
create table t_user(
 id bigint(20) PRIMARY key auto_increment comment '编号',
 user_name varchar(60) not null comment '用户名称',
 cnname varchar(60) not null comment '姓名',
 sex tinyint(3) not null comment '性别',
 mobile varchar(20) not null comment '手机号码',
 email varchar(60) not null comment '电子邮件',
 note varchar(1024) comment '备注'
);
```

2．需求

实现对用户数据的添加、删除、更新、查询。

3．创建 Maven 项目并添加依赖

（1）打开 Eclipse，创建 Maven 项目，如图 2-2 所示。

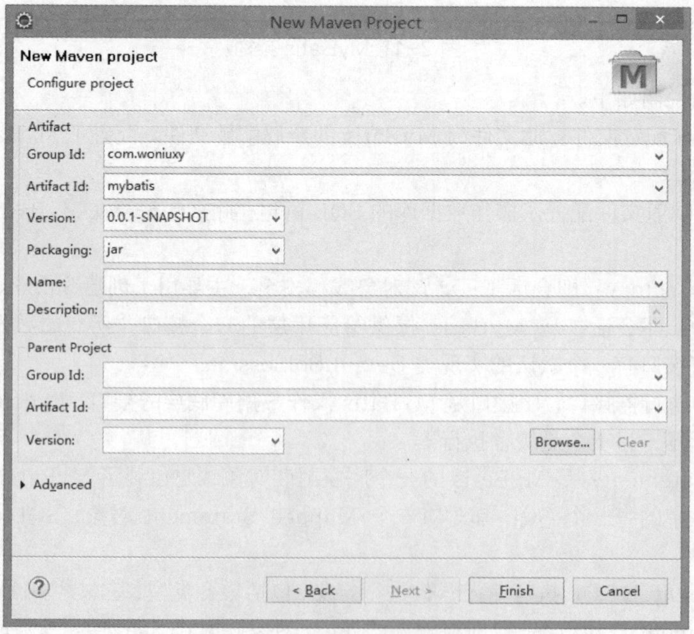

图 2-2 创建 Maven 项目

（2）添加依赖，代码如下。

```xml
<dependencies>
    <dependency>
        <groupId>org.mybatis</groupId>
        <artifactId>mybatis</artifactId>
        <version>3.4.1</version>
    </dependency>
    <dependency>
        <groupId>mysql</groupId>
        <artifactId>mysql-connector-java</artifactId>
        <version>5.1.38</version>
    </dependency>
    <dependency>
        <groupId>log4j</groupId>
        <artifactId>log4j</artifactId>
        <version>1.2.17</version>
    </dependency>
    <dependency>
        <groupId>org.slf4j</groupId>
        <artifactId>slf4j-log4j12</artifactId>
        <version>1.7.21</version>
    </dependency>
    <dependency>
        <groupId>junit</groupId>
        <artifactId>junit</artifactId>
        <version>4.12</version>
        <scope>test</scope>
    </dependency>
</dependencies>
```

4. 编写 mybatis.xml 文件

在创建好的 Maven 项目的 src/main/resources 目录中添加 mybatis.xml 文件，内容如下，用户名和密码请读者根据自己的数据库环境进行配置。

V2-2 MyBatis 环境搭建

```xml
<?xml version="1.0" encoding="utf-8" ?>
<!DOCTYPE configuration
    PUBLIC "-//mybatis.org//DTD Config 3.0//EN"
    "http://mybatis.org/dtd/mybatis-3-config.dtd">
<configuration>

    <!-- JDBC信息 -->
    <environments default="mysql">
        <environment id="mysql">
            <!-- 事务管理器 -->
            <transactionManager type="JDBC"/>
            <!-- 配置数据源 -->
            <dataSource type="POOLED">
                <property name="driver" value="org.gjt.mm.mysql.Driver"/>
                <property name="url" value="jdbc:mysql:///mybatis"/>
                <property name="username" value="用户名"/>
                <property name="password" value="密码"/>
            </dataSource>
        </environment>
    </environments>

</configuration>
```

5. 编写实体类

在 src/main/java 中创建 User.java 文件，编写用户表的实体映射文件，该类的属性需要和用户表的结构一致。

```java
public class User {
    private int id;
    private String user_name;
    private String cnname;
    private int sex;
    private String mobile;
    private String email;
    private String note;
```

6. 编写映射文件

（1）在 src/main/resources 目录中创建文件夹 mappings，如图 2-3 所示。

图 2-3　创建文件夹 mappings

（2）在 mappings 文件夹中创建 user.xml 映射文件，其内容如下。

```xml
<?xml version="1.0" encoding="utf-8" ?>
<!DOCTYPE mapper PUBLIC "-//mybatis.org//DTD Mapper 3.0//EN"
"http://mybatis.org/dtd/mybatis-3-mapper.dtd">
<mapper namespace="test">

    <!-- 根据ID查询用户的信息 -->
    <select id="queryById" parameterType="int"
        resultType="com.woniu.mybatis.entity.User">
        SELECT * from t_user where id=#{value}
    </select>

    <!-- 根据用户名进行模糊查询 -->
    <select id="queryByName" parameterType="string"
        resultType="com.woniu.mybatis.entity.User">
        SELECT * from t_user where cnname LIKE '${value}%'
    </select>

    <!-- 根据ID删除用户信息 -->
    <delete id="deleteById" parameterType="int">
        DELETE from t_user where id=#{value}
    </delete>

    <!-- 更新用户名 -->
    <update id="updateUserInfo" parameterType="com.woniu.mybatis.entity.User">
        update t_user set cnname=#{cnname} where id=#{id}
    </update>

    <!-- 添加用户信息 -->
    <insert id="saveUserInfo" parameterType="com.woniu.mybatis.entity.User">
        INSERT INTO t_user(user_name,cnname,sex,mobile,email,note)
```

```
        VALUES(#{user_name},#{cnname},#{sex},#{mobile},#{email},#{note})
    </insert>

</mapper>
```

7. 测试

（1）在 src/test/java 中编写 UserTest 测试类。

（2）通过 JUnit 简化测试，创建 SqlSession 对象。

```
public class UserTest {

    private SqlSession sqlSession;

    @Before
    public void before() throws IOException {
        //加载配置文件
        InputStream is = Resources.getResourceAsStream("mybatis.xml");
        SqlSessionFactoryBuilder builder=new SqlSessionFactoryBuilder();
        //得到SqlSessionFactory对象
        SqlSessionFactory sqlSessionFactory = builder.build(is);
        is.close();
        is=null;
        sqlSession = sqlSessionFactory.openSession();
    }

    @After
    public void after(){
        if(sqlSession!=null){
            sqlSession.close();
        }
    }
```

（3）添加用户信息。

```
@Test
    public void saveUserInfo(){
        User user=new User();
        user.setCnname("张三2");
        user.setEmail("fasdfasf");
        user.setMobile("110");
        user.setNote("fadsfad");
        user.setSex('0');
        user.setUser_name("fasdfa");

        sqlSession.insert("test.saveUserInfo", user);
        sqlSession.commit();
    }
```

（4）通过 ID 删除用户信息。

```
@Test
    public void deleteById(){
        sqlSession.delete("test.deleteById", 4);
        sqlSession.commit();
    }
```

V2-3 添加用户信息

（5）修改用户信息。

```
@Test
    public void updateUser(){
        User user=new User();
```

```
            user.setCnname("张三2");
            user.setId(1);

            int update = sqlSession.update("test.updateUserInfo", user);
            System.out.println(update);
        }
```

（6）通过 ID 查询用户信息。

```
@Test
    public void queryById(){
        try {
            /**
             * 第一个参数：执行的SQL语句的id, 即namespace.id
             * 第二个参数：传入SQL语句的参数
             */
            User user= sqlSession. selectOne ("test.queryById", 1);
            System.out.println(user.getCnname());
        } catch (Exception e) {
            e.printStackTrace();
        }
    }
```

V2-4 删除和更新用户信息

（7）通过姓名模糊查询用户信息。

```
@Test
    public void queryByName(){
        try {
            /**
             * 第一个参数：执行的SQL语句的id, 即namespace.id
             * 第二个参数：传入SQL语句的参数
             */
            List<User> users = sqlSession. selectList ("test.queryByName", "张");
            System.out.println(user.get(0).getCnname());
        } catch (Exception e) {
            e.printStackTrace();
        }
    }
```

V2-5 通过 ID 查询用户信息

通过以上案例可以看出，使用 MyBatis 操作数据库需要用户自己在映射文件中编写 SQL 语句，再通过 parameterType 指定参数的数据类型，使用 resultType 定义输出的数据类型，应用中通过 SqlSession 中的方法来实现数据库的持久化。SqlSession 中的方法如下所示。

（1）insert：用于执行添加操作，该方法需要提交事务。

（2）update：用于执行更新操作，该方法需要提交事务。

（3）delete：用于执行删除操作，该方法需要提交事务。

（4）selectOne：用于查询唯一数据，但结果由执行的 SQL 语句决定，要求真实执行的 SQL 语句的返回结果只有一条数据。

（5）selectList：用于查询返回集合类型的数据。

V2-6 selectOne 和 selectList 的区别

绑定参数过程中使用了"#""$"，这两个符号都支持绑定基本类型的数据和 POJO 类型的数据，对于 POJO 类型的数据，添加属性名即可实现数据的绑定，但对于基本类型的数据，"$" 只能用于 value 而不能用于其他数据，下面代码是错误的写法。

```
<!-- 通过姓名进行模糊查询 -->
<select id="findUserByCnname" parameterType="string" resultType="user">
    SELECT * from t_user where cnname LIKE '${dd}%'
</select>
```

V2-7 #和$的区别

2.1.3 MyBatis 动态代理开发

1. 传统的 Dao 编程方式

Java 是一门面向对象的语言，其具有封装、继承和多态的特性，但 Java 类的单继承的特性使得扩展变得相对复杂，不过可以通过接口的方式来实现多继承的特性。接口的多继承弥补了单继承的短板，提升了应用的扩展性，这就是通常所说的面向接口编程。一个接口可以有多个实现类，数据持久层对外提供统一的接口，对内具体采用何种技术实现数据的持久化并不考虑。

下面先介绍如何通过传统的面向接口的方式实现数据的持久化。

（1）编写接口，代码如下。

```java
public interface UserDao {

    void saveUserInfo(User user);

    void updateUser(User user);

    void deleteById(int id);

    User findById(int id);

    List<User> findByUser(String name);

}
```

（2）编写映射文件，内容如下。

```xml
<?xml version="1.0" encoding="utf-8" ?>
<!DOCTYPE mapper PUBLIC "-//mybatis.org//DTD Mapper 3.0//EN"
"http://mybatis.org/dtd/mybatis-3-mapper.dtd">
<mapper namespace="test">

    <!-- 根据ID查询用户的信息 -->
    <select id="queryById" parameterType="int"
        resultType="com.woniu.mybatis.entity.User">
        SELECT * from t_user where id=#{value}
    </select>

    <!-- 根据用户名进行模糊查询 -->
    <select id="queryByName" parameterType="string"
        resultType="com.woniu.mybatis.entity.User">
        SELECT * from t_user where cnname LIKE '${value}%'
    </select>

    <!-- 根据ID删除用户信息 -->
    <delete id="deleteById" parameterType="int">
        DELETE from t_user where id=#{value}
    </delete>

    <!-- 更新用户信息 -->
    <update id="updateUserInfo"
        parameterType="com.woniu.mybatis.entity.User">
        update t_user set cnname=#{cnname} where id=#{id}
    </update>

    <!-- 添加用户信息 -->
```

```xml
    <insert id="saveUserInfo"
        parameterType="com.woniu.mybatis.entity.User">
        <selectKey keyColumn="id" keyProperty="id" order="AFTER"
         resultType="int">
            SELECT LAST_INSERT_ID()
        </selectKey>
        INSERT INTO t_user(user_name,cnname,sex,mobile,email,note)
 VALUES(#{user_name},#{cnname},#{sex},#{mobile},#{email},#{note})
    </insert>

</mapper>
```

（3）编写实现类，代码如下。

```java
public class UserDaoImpl implements UserDao {

    private SqlSession sqlSession;

    public void setSqlSession(SqlSession sqlSession) {
        this.sqlSession = sqlSession;
    }

    @Override
    public void saveUserInfo(User user) {
        sqlSession.insert("test.saveUserInfo", user);
        sqlSession.commit();
        sqlSession.close();
    }

    @Override
    public void updateUser(User user) {
        sqlSession.update("test.updateUserInfo", user);
        sqlSession.commit();
        sqlSession.close();
    }

    @Override
    public void deleteById(int id) {
        sqlSession.delete("test.deleteById", id);
        sqlSession.commit();
        sqlSession.close();
    }

    @Override
    public User findById(int id) {
        User user = sqlSession.selectOne("test.queryById", id);
        sqlSession.close();
        return user;
    }

    @Override
    public List<User> findByUser(String name) {
        List<User> users = sqlSession.selectList("test.queryByName", name);
        sqlSession.close();
        return users;
    }
```

}

以上演示了在 MyBatis 中实现传统的 Dao(Data Access Object,数据访问对象)开发的全过程,可知全过程分为编写接口、编写映射文件(SQL 语句的执行单元)和编写接口的实现类,在接口的实现类中,除了实现具体的逻辑方法以外,还提供了 setSqlSession 方法来注入 SqlSession 对象。

V2-8 MyBatis 动态代理开发

2. Mapper 代理开发

前面的案例实现了数据的持久化,但其流程比较复杂,既要编写接口和接口的实现类,又要编写映射文件(SQL 语句的执行单元),这无形中加大了开发人员的工作量。是否存在更简洁的方式呢?答案是肯定的,下面将介绍如何使用 MyBatis 提供的代理开发来简化数据的持久化的实现。

由于本书不是专门介绍 MyBatis 的,所以读者若想了解其动态代理如何实现,可查看相关书籍,这里只做简单的使用层面的介绍。MyBatis 的动态代理开发仅需要开发人员提供相应的接口和映射文件即可,但编写上必须满足以下几条规则。

(1)映射文件中的 namespace 需要和接口的全限定名保持一致。
(2)映射文件中 statement 语句的 ID 必须和接口的方法名保持一致。
(3)映射文件中 statement 的 parameterType 必须和接口方法的参数类型保持一致。
(4)映射文件中 statement 的 resultType 必须和接口方法的返回值保持一致。

在开发过程中,必须严格遵守以上规则。下面通过简单的案例来讲解具体操作的实现。

(1)编写代理接口,代码如下。

```java
public interface UserMapper {

    void saveUserInfo(User user);

    void updateUser(User user);

    void deleteById(int id);

    User queryById(int id);

    List<User> queryByName(String name);
}
```

(2)编写映射文件,内容如下。

```xml
<?xml version="1.0" encoding="utf-8" ?>
<!DOCTYPE mapper PUBLIC "-//mybatis.org//DTD Mapper 3.0//EN"
"http://mybatis.org/dtd/mybatis-3-mapper.dtd">
<mapper namespace="com.woniu.mybatis.mapper.UserMapper">

    <!-- 根据ID查询用户的信息 -->
    <select id="queryById" parameterType="int"
        resultType="com.woniu.mybatis.entity.User">
        select * FROM t_user where id=#{value}
    </select>

    <!-- 根据用户名进行模糊查询 -->
    <select id="queryByName" parameterType="string"
        resultType="com.woniu.mybatis.entity.User">
        SELECT * from t_user where cnname LIKE '${value}%'
    </select>

    <!-- 根据ID删除用户信息 -->
```

```xml
<delete id="deleteById" parameterType="int">
    DELETE from t_user where id=#{value}
</delete>

<!-- 更新用户信息 -->
<update id="updateUser"
    parameterType="com.woniu.mybatis.entity.User">
    update t_user set cnname=#{cnname} where id=#{id}
</update>

<!-- 添加用户信息 -->
<insert id="saveUserInfo"
    parameterType="com.woniu.mybatis.entity.User">
    <selectKey keyColumn="id" keyProperty="id" order="AFTER"
     resultType="int">
        SELECT LAST_INSERT_ID()
    </selectKey>
    INSERT INTO t_user(user_name,cnname,sex,mobile,email,note) VALUES(#{user_name},#{cnname},#{sex},#{mobile},#{email},#{note})
</insert>

</mapper>
```

（3）进行测试，代码如下。

```java
@Test
public void save(){
    //通过SqlSession中的方法获取Mapper的代理对象
    UserMapper userMapper = sqlSession.getMapper(UserMapper.class);
    User user=new User();
    user.setCnname("fasdfasdf");
    user.setUser_name("fadsf");
    user.setEmail("fasdf");
    user.setSex('1');
    user.setMobile("fasdf");
    user.setNote1("fadsf");
    //调用Mapper中的方法完成数据的持久化
    userMapper.saveUserInfo(user);

    sqlSession.commit();
}
```

以上演示了如何使用 MyBatis 的动态代理开发来实现数据的持久化，相对于传统的 Dao 开发，其省去了实现类的编写，这也是目前使用最多的方式，是 MyBatis 官网推荐使用的方式。

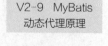

V2-9 MyBatis 动态代理原理

2.1.4 MyBatis 关系映射

通过前面的讲解，相信读者应该掌握了使用 MyBatis 完成数据持久化的方法，以上操作的是关系型数据库，而关系型数据中表与表之间是存在着一对一、一对多或多对多关系的。通过前面的学习可知 MyBatis 是一个优秀的 ORM（Object/Relational Mapping，对象-关系映射）框架，通过对象的方式来展现关系型数据，那么在 MyBatis 中是如何完成这种关系映射的呢？

1. 通过 resultMap 自定义映射

前面在执行查询的 statement 语句中，使用 resultType 来指定返回的结果类型，将 ResultSet 中的数据绑定到相信的数据类型中，如 POJO（Plain Ordinary Java Object，简单 Java 对象）对象中，但前提是实体类的属性名需要和数据库表的字段名一致。然而，在实际开发中难免会遇到不一致的情况，在这些情况下，如何实现数据的绑定呢？MyBatis 为开发人员提供了 resultMap 来完成自定义映射。

（1）修改实体类，使其属性名和数据库表列名不一致。

```java
public class User {
    private int id;
    private String user_name;
    private String cnname;
    private char sex;
    private String mobile;
    private String email;
    private String note1;
```

（2）自定义映射规则。在映射文件中使用 resultMap 定义映射的规则，完成 ResultSet 中每行数据的绑定。

```xml
<!--
    type表示返回的数据类型，可以定义别名
    id表示唯一标志
-->
<resultMap type="user" id="mappingUserInfo">
    <!-- 映射主键
        column：查询得到的列名（别名）
        property：映射到该实体中的某个属性的属性名
        javaType：Java中的数据类型
        jdbcType：数据库中该列的数据类型
    -->
    <id column="uId" property="id" javaType="int"/>
    <!-- 映射普通列 -->
    <result column="user_name" property="user_name" javaType="string"/>
    <result column="cName" property="cnname" javaType="string"/>
    <result column="mobile" property="mobile" javaType="string"/>
    <result column="email" property="email" javaType="string"/>
    <result column="sex" property="sex" jdbcType="TINYINT"/>
    <result column="note" property="note1" javaType="string"/>
</resultMap>
```

（3）引用映射规则，代码如下。

```xml
<select id="queryById" parameterType="int"
        resultMap="mappingUserInfo">
    select * FROM t_user where id=#{value}
</select>
```

（4）进行测试，代码如下。

```java
@Test
public void queryById(){
    UserMapper userMapper = sqlSession.getMapper(UserMapper.class);
    User user = userMapper.queryById(1);
    System.out.println(user.getUser_name());
}
```

2. 关联 association

在实际开发中，或多或少地会遇到一个实体类中关联了另一个实体类的情况。例如，获取学生信息的同时希望获取该学生的学生证信息，即学生和学生证存在一对一关系。MyBatis 中提供了 association 节点来处理这种一个实体类关联另一个实体类的情况，实现了另一个实体类直接映射到实体类的复合属性中。

V2-10 resultMap 自定义映射

（1）编写实体类。编写学生及学生证实体类的代码如下。

```java
public class Student{  //学生实体类

    private int id;
    private String cnname;
    private char sex;
    private String note;
}

public class IdCard {  //学生证实体类

    private int id;
    private int student_id; //学生ID
    private String natives;
    private Date issue_date;
    private Date end_date;
    private String note;
}
```

（2）添加关联属性。在学生实体类中添加学生证实体类的关联属性，用于将其关联的信息绑定到该属性中。

```java
public class Student{

    private IdCard idCard;
}
```

（3）编写 StudentMapper 接口，代码如下。

```java
public interface StudentMapper {

    Student findById(int id);
}
```

（4）在映射文件中配置关联映射。

```xml
<resultMap type="student" id="onebyone">
    <id column="id" property="id"/>
    <result column="cnname" property="cnname"/>
    <result column="sex" property="sex"/>
    <result column="note" property="note"/>
    <!--
        一对一的映射：映射学生证表中的数据，将它转换为Java中IdCard类型的数据
        property：实体类中的属性名
        javaType：数据的类型
    -->
    <association property="idCard" javaType="idCard">
        <!-- 映射主键 -->
        <id column="tss_id" property="id"/>
        <result column="native" property="natives"/>
        <result column="issue_date" property="issue_date"/>
        <result column="end_date" property="end_date"/>
        <result column="id" property="student_id"/>
    </association>
</resultMap>
```

（5）引用映射，代码如下。

```xml
<select id="findById" parameterType="int" resultMap="onebyone">
    SELECT
        ts.*, tss.id AS tss_id,
        tss.native native,
```

```
        tss.issue_date,
        tss.end_date
    FROM
        t_student ts
    INNER JOIN t_student_selfcard tss ON ts.id = tss.student_id
    WHERE
        ts.id = #{id}
</select>
```

（6）进行测试，代码如下。

```
@Test
public void oneByOne(){
    StudentMapper studentMapper =
    sqlSession.getMapper(StudentMapper.class);
    Student student = studentMapper.findById(1);
    System.out.println(student.getIdCard().getNatives());
}
```

3. 集合映射 collection

通过以上的讲解，可以了解到在 MyBatis 中通过 association 节点将关联的属性映射到实体类的复合属性中，但该复合属性只能存在于单个 POJO 对象中，而在实际开发中会遇到将多条数据映射到实体类对象的集合属性中的情况。MyBatis 提供了 collection 节点来处理这类问题，从其命名上可以看出这个节点是 MyBatis 为开发人员提供的用于映射集合类型数据的接口。下面通过获取学生的体检信息来讲解 collection 的应用。

V2-11 association 一对一关系映射

（1）编写实体类，代码如下。

```java
public class HealthInfo {

    private int id;
    private int student_id;
    private Date check_date;
    private String heart;
    private String liver;
    private String spleen;
    private String lung;
    private String kidney;
    private String prostate;
    private String note;
}
```

（2）添加关联属性。在学生实体类中添加集合属性，用于映射体检信息，代码如下。

```java
public class Student{

    private List<HealthInfo> healthInfos;
}
```

（3）修改 StudentMapper 接口，添加如下方法。

```java
public interface StudentMapper {

    Student findHealthInfo(int id);
}
```

（4）在映射文件中配置关联映射，内容如下。

```xml
<resultMap type="student" id="onetoMany">
    <id column="id" property="id"/>
    <result column="cnname" property="cnname"/>
    <result column="sex" property="sex"/>
```

```xml
        <result column="note" property="note"/>
    <!-- 多个结果集的映射
        property：属性的名称
        ofType：该集合中存在的数据类型
     -->
    <collection property="healthInfos" ofType="HealthInfo">
        <id column="tshId" property="id"/>
        <result column="id" property="student_id"/>
        <result column="check_date" property="check_date"/>
        <result column="heart" property="heart"/>
        <result column="liver" property="liver"/>
    </collection>
</resultMap>
```

（5）引用映射，代码如下。

```xml
<select id="findHealthInfo" resultMap="onetoMany">
    SELECT
        ts.*,
      tsh.id tshId,
      tsh.check_date,
      tsh.heart,
      tsh.liver,
      tsh.spleen,
      tsh.lung,
      tsh.kidney,
      tsh.prostate,
      tsh.note as notetsh
    FROM
      t_student ts
    INNER JOIN t_student_health_male tsh
    on
      tsh.student_id = ts.id where ts.id=#{id}
</select>
```

（6）进行测试，代码如下。

```java
@Test
public void oneByMany(){
    StudentMapper studentMapper = sqlSession.getMapper(StudentMapper.class);
    Student student = studentMapper.findHealthInfo(1);
    System.out.println(student.getHealthInfos().get(0).getHeart());
}
```

通过本节的讲解，读者可以解决开发中的大部分问题，但由于本书不专门讲解 MyBatis，故对 MyBatis 的一些高级应用，如动态 SQL、延迟加载和缓存等，不做过多介绍，有兴趣的读者可以查阅相关资料，也可以登录蜗牛学院官网查看相应的视频。

V2-12 collection 一对多关系映射

2.2 Spring 概述

2.2.1 了解 Spring

Spring 是由 Rod Johnson（罗宾·约翰逊）领导开发的开源的、免费的、民间的框架。Rod Johnson 曾是 Servlet 2.4 规范专家组的成员，但他对 EJB（Enterprise JavaBean，企业级 JavaBean）等官方标准表示怀疑，认为过于复杂、臃肿的重量级官方框架并不能为解决问题提供便利。为此，2002 年他编写

2. 添加日志文件
在 src/main/resources 目录中创建 log4j.properties 文件，其内容如下。

```
### direct log messages to stdout ###
log4j.appender.stdout=org.apache.log4j.ConsoleAppender
log4j.appender.stdout.Target=System.out
log4j.appender.stdout.layout=org.apache.log4j.PatternLayout
log4j.appender.stdout.layout.ConversionPattern=%d{ABSOLUTE} %5p %c{1}:%L - %m%n

log4j.rootLogger=debug, stdout
```

3. 添加配置文件
在 src/main/resources 目录中创建 beans.xml 文件，其内容如下。

```xml
<?xml version="1.0" encoding="UTF-8"?>
<beans xmlns="http://www.springframework.org/schema/beans"
    xmlns:xsi="http://www.w3.org/2001/XMLSchema-instance"
    xmlns:context="http://www.springframework.org/schema/context"
    xsi:schemaLocation="http://www.springframework.org/schema/beans
        http://www.springframework.org/schema/beans/spring-beans-4.2.xsd
        http://www.springframework.org/schema/context
       http://www.springframework.org/schema/context/spring-context-4.2.xsd">

</beans>
```

4. 初始化 IoC 容器
前面已经提到过，Spring 通过 BeanFactory 对象使用控制反转模式将应用程序的配置和依赖性规范与实际的应用程序代码分开，但 BeanFactory 是 Spring 底层的实现，在实际开发中一般使用该接口的 ApplicationContext 子接口，该接口是 Spring 提供给应用层使用的，该接口存在 4 个常用的实现类，针对不同环境下的 Spring 应用进行配置。

（1）ClassPathXmlApplicationContext：基于类路径的 XML 文件配置。

（2）FileSystemXmlApplicationContext：基于文件系统的 XML 文件配置。

（3）AnnotationConfigApplicationContext：基于 JavaConfig 的配置。

（4）WebApplicationContext：针对 Web 环境的配置。

这里使用基于类路径的 ClassPathXmlApplicationContext 对象来初始化 Spring 容器，在 test 包中创建 InitSpringContainer 类，其内容如下。

```java
public class InitSpringContainer {

    @Test
    public void init(){
        ApplicationContext act=new ClassPathXmlApplicationContext("beans.xml");
    }

}
```

5. 在 Spring 中装配 Bean
前面已经成功地初始化了 IoC 容器，那么在 IoC 容器中如何对 Bean 进行装配呢？Spring 中提供了 BeanDefinition 接口对象对 Bean 进行封装，该对象中定义了 Bean 的元数据信息，封装了 Bean 的全限定名、Bean 的行为及声明周期等相关信息。所以要使用 Spring 创建对象，必须对相应的元数据信息进行配置。

（1）创建 UserAction 对象，代码如下。

```java
public class UserAction {
}
```

（2）在 Spring 中装配对象，代码如下。

```
<!--
    id：Bean的唯一标志
    class：Bean的全限定名
-->
<bean id="userAction" class="com.woniu.spring.action.UserAction"/>
```

6. 装配对象间的关系

在配置文件中添加 Bean 的元数据信息，Spring 在初始化的时候会根据这些元数据信息进行对象的创建，即应用本身不负责对象的创建，而交由外部容器进行处理。然而，在应用中，对象与对象势必会存在某种关系，Spring 通过依赖注入的方式对这种关系进行了处理。常用的依赖注入方式有以下两种。

V2-15　Spring
构造函数初始化对象

（1）构造函数注入：通过构造函数注入依赖的对象。

（2）set 方法注入：通过 set 方法注入依赖的对象。

① 创建 UserService 并在 Spring 中进行装配，代码如下。

```java
public class UserService {

}
```

② 修改 UserAction，添加 UserService 的 set 方法，代码如下。

```java
public class UserAction {

    private UserService userService;

    public void setUserService(UserService userService) {
        this.userService = userService;
    }
}
```

③ 装配关系，内容如下。

```xml
<bean id="userService" class="com.woniu.spring.service.UserService"/>

<bean id="userAction" class="com.woniu.spring.action.UserAction">
    <!-- 通过set方法注入 -->
    <property name="userService" ref="userService"/>
</bean>
```

通过前面的讲解，相信读者对 Spring 中的 IoC 有了一定的了解，当然，这仅是从应用方面讲解了如何对 Bean 进行装配以及关联对象如何进行装配。Spring 中除了可装配自定义的对象以外，也支持集合类型及基本数据类型的装配，对这部分内容有兴趣的读者可以查阅相关资料，结合以上案例理解 Spring 是如何对 Bean 进行管理的。

2.2.3　Spring 的 AOP 编程

V2-16　Spring
依赖注入

Java 开发人员对面向对象程序设计（Object Oriented Programming，OOP）很熟悉，那么 Spring 中的 AOP 具体代表什么含义呢？先思考一个问题：有时需要对系统中已有的某些业务做日志记录，如考虑为已有的支付系统的支付业务添加记录支付日志的功能。这对于已有支付系统而言可能相当复杂，因为支付的功能可能由自己实现，也可能采用第三方支付平台，从而无法进行统一处理（后者无法得到源码）。面对这样的支付系统，该如何解决呢？

1. 传统解决方案

（1）日志部分提取公共类 LogUtils，定义 logPayBegin 方法，用于记录支付开始日志，定义 logPayEnd 方法，用于记录支付结果。

（2）对于支付部分，定义 IPayService 接口并定义支付方法 pay，定义两个实现——PointPayService（表示积分支付）和 RMBPayService（表示人民币支付），并在每个支付中实现支付逻辑和日志记录。

以上设计了一个可以复用的接口，但有没有更好的解决方案呢？如果对积分支付方式添加统计功能，如在支付时记录用户总积分数、当前消费的积分数，应该如何做呢？可以直接修改源代码添加日志记录，但这样完全违背了面向对象最重要的原则之一：开闭原则（对扩展开放，对修改关闭）。

2. 更好的解决方案

由于支付组件中使用了日志组件，即日志模块横切于支付组件，在传统程序设计中很难将日志组件分离出来，即不耦合支付组件，因此面向切面编程诞生了，它能分离组件，使组件完全不耦合。

采用面向切面编程后，支付组件只负责处理支付的逻辑，代码中不再有日志组件的任何内容；所有日志相关的内容提取到一个切面中，AOP 实现者会在合适的时候将日志功能织入到支付组件中，从而完全解耦支付组件和日志组件。

前面出现了一个词语——横切，这和 AOP 有什么联系呢？AOP 是通过预编译方式和运行期动态代理实现程序功能的一种技术，是一种横向的编程方式，用于解决 OOP 这种纵向编程无法解决的问题。AOP 中有很多术语，这里先做简单介绍。

（1）连接点（Jointpoint）：表示需要在程序中插入横切关注点的扩展点。连接点可能是类初始化、方法执行、方法调用、字段调用或处理异常等连接点，Spring 只支持方法执行连接点，在 AOP 中表示为"在哪里做"。

（2）切入点（Pointcut）：是连接点的集合，Spring 支持 Perl 5 正则表达式和 AspectJ 切入点模式，默认使用 AspectJ 语法。

（3）增强（Advice）：表示在连接点上执行的行为，提供了 AOP 中在切入点所选择的连接点处扩展现有行为的手段，包括前置增强（before advice）、后置增强（after advice）和环绕增强（around advice），在 Spring 中通过代理模式实现 AOP，并通过过滤器模式以环绕连接点的过滤器链织入增强，在 AOP 中表示为"做什么"。

（4）切面（Aspect）：是横切关注点的模块化，例如，前面提到的日志组件可以认为是增强、引入和切入点的组合。切面可以在 Spring 中使用 Schema 和@AspectJ 的方式进行组织实现，在 AOP 中表示为"在哪儿做和做什么集合"。

（5）内部类型声明（Inter-Type Declaration）：也称引入，为已有的类添加额外的新字段或方法，Spring 允许引入新的接口（必须对应一个实现）到所有被代理对象（目标对象），在 AOP 中表示为"做什么（引入什么）"。

（6）目标对象（Target Object）：表示需要被织入横切关注点的对象，即该对象是切入点选择的对象、需要被增强的对象，因此也可称为"被增强对象"；由于 Spring AOP 通过代理模式实现，因此这个对象永远是被代理对象，在 AOP 中表示为"对谁做"。

（7）AOP 代理（AOP Proxy）：AOP 框架使用代理模式创建对象，从而实现在连接点处插入增强（即应用切面），即通过代理来对目标对象应用切面。在 Spring 中，AOP 代理可以用 JDK 动态代理或 CGLIB 代理实现，而通过过滤器模型应用切面。

（8）织入（Weaving）：织入是一个过程，是将切面应用到目标对象从而创建出 AOP 代理对象的过程，织入可以在编译期、类装载期、运行期进行。

通过上面的介绍，相信读者对 AOP 有了简单的了解，接下来来了解如何通过 Spring 提供的 AOP 完成以上支付系统中日志的记录。

V2-17 Spring AOP 及其实现原理

1. 创建项目并添加依赖

```
<dependency>
    <groupId>org.springframework</groupId>
    <artifactId>spring-core</artifactId>
    <version>4.3.13.RELEASE</version>
</dependency>
```

```xml
<dependency>
    <groupId>org.springframework</groupId>
    <artifactId>spring-beans</artifactId>
    <version>4.3.13.RELEASE</version>
</dependency>
<dependency>
    <groupId>org.springframework</groupId>
    <artifactId>spring-context</artifactId>
    <version>4.3.13.RELEASE</version>
</dependency>
<dependency>
    <groupId>org.springframework</groupId>
    <artifactId>spring-aop</artifactId>
    <version>4.3.13.RELEASE</version>
</dependency>
<dependency>
    <groupId>org.springframework</groupId>
    <artifactId>spring-aspects</artifactId>
    <version>4.3.13.RELEASE</version>
</dependency>
```

2. 编写支付接口及实现类

```java
public interface IPayService {

    /**
     * 积分支付
     * @param money
     * @return
     */
    String payByPoint(double money);

    /**
     * 人民币支付
     * @param money
     * @return
     */
    String payRMB(double money);

}


@Service("IPayService")
public class IPayServiceImpl implements IPayService {

    private static double total=100000000;
    private static double totalP=100000000;

    @Override
    public String payByPoint(double money) {
        BigDecimal decimal=new BigDecimal(totalP);
        if(decimal.subtract(new BigDecimal(money)).compareTo(BigDecimal.ZERO)>=0){
            totalP=decimal.subtract(new BigDecimal(money)).floatValue();
            return "success";
        }else{
            return "fail";
        }
    }
```

```java
    @Override
    public String payRMB(double money) {
        BigDecimal decimal=new BigDecimal(total);
        if(decimal.subtract(new BigDecimal(money)).compareTo(BigDecimal.ZERO)>=0){
            total=decimal.subtract(new BigDecimal(money)).floatValue();
            return "success";
        }else{
            return "fail";
        }
    }
}
```

3. 编写增强

```java
public class LoggerAdvice {

    private static final Logger LOGGER=Logger.getLogger(LoggerAdvice.class);

    public void before(){
        System.out.println("在方法执行之前执行");
    }

    public void after(){
        System.out.println("在方法执行之后执行");
    }

    /**
     * 回调函数
     * @param joinPoint
     * @return
     */
    public Object around(ProceedingJoinPoint joinPoint){
        /*joinPoint.getArgs(); //得到实际参数
        joinPoint.getTarget(); //得到目标对象
        joinPoint.getSignature(); //得到方法签名
        */
        try {
            //记录支付了多少钱
            Object money = joinPoint.getArgs()[0];
            Object result = joinPoint.proceed(); //完成对目标对象方法的调用
            //支付成功
            LOGGER.debug("支付了："+money+"  结果为："+result);

            return result;
        } catch (Throwable e) {
            LOGGER.debug("支付失败"+e.getMessage());
        }
        return null;
    }

}
```

4. 在 Spring 中进行装配

```xml
<context:component-scan base-package="com.woniuxy.spring.service.impl" />

<!-- 配置增强 -->
```

```xml
<bean id="logger" class="com.woniuxy.spring.advice.LoggerAdvice"/>

<!-- 配置AOP -->
<aop:config>
    <!-- 切入点 -->
    <aop:pointcut expression="execution(* com.woniuxy.spring.service.impl.
IPayServiceImpl.*(..))" id="pointCut"/>

    <!--切面：增强和切入点 -->
    <aop:aspect ref="logger">
        <aop:after method="after" pointcut-ref="pointCut"/>
        <aop:before method="before" pointcut-ref="pointCut"/>
        <aop:around method="around" pointcut-ref="pointCut"/>
    </aop:aspect>
</aop:config>
```

5. 测试

```java
@RunWith(SpringJUnit4ClassRunner.class)
@ContextConfiguration(locations="classpath:beans.xml")
public class IPServiceTest {

    @Resource
    private IPayService iPayService;

    @Test
    public void pay(){
        String result = iPayService.payByPoint(1000000000);
        System.out.println(result);
    }

}
```

通过测试可以看到，在执行支付时，相应的日志信息也被记录下来了，在真实项目中，这些日志数据会持久化到数据库中。

2.3 Spring MVC 概述

2.3.1 Spring MVC 简介

Spring MVC 是一种基于 Java 的、实现 Web MVC 设计模式的请求驱动类型的轻量级 Web 框架，使用了 MVC 架构模式的思想，对 Web 层进行解耦。基于请求驱动指的就是使用请求-响应模型，框架的目的是帮助用户简化开发，Spring MVC 也用于简化日常 Web 开发。Spring MVC 和 Struts 2 都属于表现层框架，Spring MVC 是 Spring 框架的一部分，这可以从 2.2 节提到的 Spring 架构中看出来。

从 Spring 官网给出的架构图中可知，Web 模块中的 Web 和 Servlet 用于集成 Web 容器及 Spring MVC 的具体实现。前面提到过 Spring MVC 是基于请求驱动的 Web 框架，下面就来了解 Spring MVC 的核心组件，如图 2-5 所示。

从图 2-5 可以看出，由用户发起请求交给 Spring MVC 的中央控制器，中央控制器委托处理器映射器根据实际请求的 URL 地址查找到对应的处理器并返回一个 HandlerExecutionChain 对象给中央控制器，得到该对象后再由中央控制器委托处理器适配器完成处理器的调用并返回一个 ModelAndView 对象，中央控制器得到 ModelAndView 对象后再交给视图解析器对模型和视图进行渲染，将渲染之后的视图作为响应发送给用户。这个流程（一个请求一个响应）就是请求驱动，所以 Spring MVC 是围绕请求来处理的。

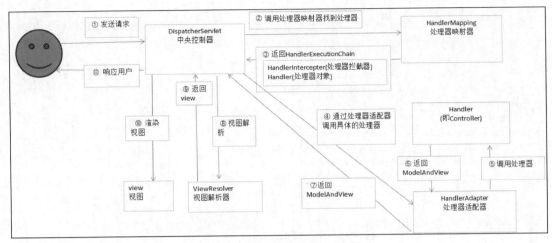

图 2-5　Spring MVC 的核心组件

这里提到了一些组件，如中央控制器、处理器适配器、处理器映射器、处理器、视图解析器，这些组件主要作用是什么？

（1）中央控制器：整个流程控制的中心，主要负责过滤用户请求，并将请求分配给相应的组件进行处理。

（2）处理器适配器：主要负责完成对处理器的调用，得到 ModelAndView 对象，其为适配器模式的应用，用于实现对不同处理器的支持。

（3）处理器映射器：主要负责根据请求的 URL 地址查找到相应的处理器。

（4）处理器：处理用户请求的真实逻辑实现部分，也称后端控制器。

（5）视图解析器：负责将处理结果生成视图，其先根据逻辑视图名解析为真实的物理视图，再生成视图对象，最后对视图进行渲染并将处理结果通过页面展示给用户。Spring MVC 框架提供了很多视图类型，包括 JstlView、FreeMarkerView、pdfView 等。

2.3.2　Spring MVC 请求处理

由前面的讲解可知，Spring MVC 主要通过处理器来处理用户的请求，下面以用户登录为例来讲解 Spring MVC 如何处理用户的请求。

V2-19　Spring MVC 介绍及环境搭建

1. 创建 Web 项目并添加依赖

```xml
<dependencies>
    <!-- Spring的核心 -->
    <dependency>
        <groupId>org.springframework</groupId>
        <artifactId>spring-beans</artifactId>
        <version>4.3.5.RELEASE</version>
    </dependency>
    <dependency>
        <groupId>org.springframework</groupId>
        <artifactId>spring-core</artifactId>
        <version>4.3.5.RELEASE</version>
    </dependency>
    <dependency>
        <groupId>org.springframework</groupId>
        <artifactId>spring-context</artifactId>
        <version>4.3.5.RELEASE</version>
    </dependency>
```

```xml
        <!-- Spring和Web容器整合的包 -->
        <dependency>
            <groupId>org.springframework</groupId>
            <artifactId>spring-web</artifactId>
            <version>4.3.5.RELEASE</version>
        </dependency>
        <!-- Spring MVC -->
        <dependency>
            <groupId>org.springframework</groupId>
            <artifactId>spring-webmvc</artifactId>
            <version>4.3.5.RELEASE</version>
        </dependency>
        <!-- servlet -->
        <dependency>
            <groupId>javax.servlet</groupId>
            <artifactId>javax.servlet-api</artifactId>
            <version>3.1.0</version>
        </dependency>
        <dependency>
            <groupId>javax.servlet.jsp</groupId>
            <artifactId>jsp-api</artifactId>
            <version>2.2.1-b03</version>
        </dependency>
        <dependency>
            <groupId>jstl</groupId>
            <artifactId>jstl</artifactId>
            <version>1.2</version>
        </dependency>
        <dependency>
            <groupId>log4j</groupId>
            <artifactId>log4j</artifactId>
            <version>1.2.17</version>
        </dependency>
        <dependency>
            <groupId>org.slf4j</groupId>
            <artifactId>slf4j-log4j12</artifactId>
            <version>1.7.21</version>
        </dependency>
        <dependency>
            <groupId>junit</groupId>
            <artifactId>junit</artifactId>
            <version>4.12</version>
            <scope>test</scope>
        </dependency>
        <dependency>
            <groupId>org.springframework</groupId>
            <artifactId>spring-test</artifactId>
            <version>4.3.5.RELEASE</version>
            <scope>test</scope>
        </dependency>
</dependencies>
```

2. 配置前端控制器

```xml
<?xml version="1.0" encoding="UTF-8"?>
<web-app xmlns:xsi="http://www.w3.org/2001/XMLSchema-instance"
    xmlns="http://java.sun.com/xml/ns/javaee"
```

```xml
    xsi:schemaLocation="http://java.sun.com/xml/ns/javaee
http://java.sun.com/xml/ns/javaee/web-app_3_0.xsd"
    id="WebApp_ID" version="3.0" >

    <servlet>
        <servlet-name>SpringMVC</servlet-name>
        <servlet-class>org.springframework.web.servlet.DispatcherServlet
</servlet-class>
        <!-- 指定Spring MVC的配置文件 -->
        <init-param>
            <param-name>contextConfigLocation</param-name>
            <param-value>classpath:SpringMVC.xml</param-value>
        </init-param>
        <load-on-startup>0</load-on-startup>
    </servlet>
    <servlet-mapping>
        <servlet-name>SpringMVC</servlet-name>
        <url-pattern>*.do</url-pattern>
    </servlet-mapping>

</web-app>
```

3. 编写 Spring MVC 的配置文件

```xml
<beans xmlns="http://www.springframework.org/schema/beans"
    xmlns:xsi="http://www.w3.org/2001/XMLSchema-instance"
    xmlns:mvc="http://www.springframework.org/schema/mvc"
    xmlns:context="http://www.springframework.org/schema/context"
    xsi:schemaLocation="http://www.springframework.org/schema/beans
    http://www.springframework.org/schema/beans/spring-beans-4.1.xsd
    http://www.springframework.org/schema/context
    http://www.springframework.org/schema/context/spring-context-4.1.xsd
    http://www.springframework.org/schema/mvc
    http://www.springframework.org/schema/mvc/spring-mvc-4.1.xsd">
    <!-- 处理器映射器 -->
    <!-- 将使用handler的Bean的name属性作为URL地址 -->
    <bean class="org.springframework.web.servlet.handler.BeanNameUrlHandlerMapping">
    </bean>

    <!-- 处理器适配器:handler的编写规则 -->
    <bean class="org.springframework.web.servlet.mvc.SimpleControllerHandlerAdapter">
    </bean>

    <!-- 视图解析器 -->
    <bean class="org.springframework.web.servlet.view.InternalResourceViewResolver">
        <!-- 视图文件的前缀 -->
        <property name="prefix" value="/WEB-INF/views/"/>
        <!-- 指定视图文件的后缀 -->
        <property name="suffix" value=".jsp"/>
    </bean>

</beans>
```

4. 编写登录界面

```jsp
<%@ page language="java" contentType="text/html; charset=UTF-8"
    pageEncoding="UTF-8"%>
<!DOCTYPE html>
<html>
```

```html
<head>
<meta http-equiv="Content-Type" content="text/html; charset=UTF-8">
<title>Insert title here</title>
</head>
<body>
    <div align="center">
        <form action="login.do" method="post">
            <table>
                <tr>
                    <td>用户名：</td>
                    <td><input type="text" name="userName"></td>
                </tr>
                <tr>
                    <td>密码：</td>
                    <td><input type="password" name="password"></td>
                </tr>
                <tr>
                    <td colspan="2" align="center">
                        <input type="submit" value="登录">
                        <input type="reset" value="重置">
                    </td>
                </tr>
            </table>
        </form>
    </div>
</body>
</html>
```

5. 编写处理器

```java
public class UserController implements Controller {

    @Override
    public ModelAndView handleRequest(HttpServletRequest request, HttpServletResponse response) throws Exception {
        ModelAndView modelAndView=new ModelAndView();
        String userName = request.getParameter("userName");
        String password = request.getParameter("password");
        if("admin".equals(password)&&"admin".equals(userName)){
            modelAndView.setViewName("success");
        }else{
            modelAndView.setViewName("fail");
        }
        return modelAndView;
    }
}
```

6. 配置处理器

```xml
<bean name="/login.do" class="com.woniu.springmvc.handler.UserController"/>
```

7. 部署测试

部署项目到 Web 容器中，输入用户名和密码进行测试，看能否跳转到登录成功界面。

8. 详解

回顾整个流程，相对于之前的 Servlet 开发，这里将处理器的配置换到了 Spring MVC 的配置文件中，视图的返回也只是返回一个 ModelAndView 对象即可。那么

V2-20 Spring MVC 用户登录

Spring MVC 的配置文件中到底配置了哪些信息呢？

（1）处理器映射器

HandlerMapping 是 Spring MVC 中提供的一个接口，主要用于根据 Request 找到对应的 handler 处理器及 Interceptor（拦截器），将它们封装在 HandlerExecutionChain 对象中并交给中央控制器。如前面配置的 BeanNameUrlHandlerMapping 用做 URL 与 Bean 名称的映射，映射成功的 Bean 就是此处的处理器。Spring MVC 除了提供基于 Bean 的名称的处理器映射器以外，还提供了以下实现类。

① BeanNameUrlHandlerMapping：根据请求的 URL 与 Spring 容器中定义的 Bean 的 name 进行匹配，从而从 Spring 容器中找到 Bean 实例。

② SimpleUrlHandlerMapping：是 BeanNameUrlHandlerMapping 的增强版本，它可以对 URL 和处理器 Bean 的 ID 进行统一映射配置。

③ RequestMappingHandlerMapping：是基于注解的处理器映射器的实现，具体使用方法将在 2.3.3 节中介绍。

（2）处理器适配器

Spring MVC 中提供的处理器适配器决定了处理器功能处理方法的调用，处理器适配器会根据适配的结果调用真正的处理器的功能处理方法，完成功能处理后会返回 ModelAndView 对象（包含模型数据、逻辑视图名）。处理器适配器也是 Spring MVC 提供的接口，但它的存在会将处理器包装为适配器，从而支持不同类型的处理器，此即适配器设计模式的应用。Spring MVC 为用户提供了以下常用的实现类。

① SimpleControllerHandlerAdapter：简单控制器处理器适配器，规定所有实现了 org.springframework.web.servlet.mvc.Controller 接口的 Bean 通过此适配器进行适配、执行。

② HttpRequestHandlerAdapter：HTTP 请求处理器适配器，规定所有实现了 org.springframework.web.HttpRequestHandler 接口的 Bean 通过此适配器进行适配、执行。

③ RequestMappingHandlerAdapter：基于注解的处理器适配器的实现。

（3）视图解析器

视图解析器是 Spring MVC 中提供的一个接口，用于将处理器放回的 ModelAndView 对象中的逻辑视图名转换成真实的物理视图并完成对数据的渲染，同时实现对不同视图引擎的支持。例如，前面配置的 InternalResourceViewResolver 主要针对使用 JSP 和 JSTL 作为视图处理时的实现，除了该实现类以外，Spring MVC 还内置了很多实现类，这里列举几个比较常用的实现类。

V2-21 Spring MVC 处理器适配器和映射器

① InternalResourceViewResolver：支持使用 JSP 作为模板引擎。

② FreeMarkerViewResolver：支持使用 FreeMarker 作为模板引擎。

③ GroovyMarkupViewResolver：支持使用 GroovyMarkup 作为模板引擎。

（4）处理器

处理器作为开发人员最直接接触的组件，用于整个处理流程中具体业务的实现，并对用户请求进行精确处理，获取用户输入的数据。在 Spring MVC 中，通过属性编辑器可以将请求的参数自动绑定到方法参数上，从而简化数据获取。Spring MVC 为开发人员提供了以下几种方式来获取用户输入的数据。

① 原生的 ServletAPI，如 HttpServletRequest、HttpServletResponse、HttpSession 对象。

② 基本类型数据，只要方法参数名和提供参数名一致即可。

③ POJO 对象。

④ Model、ModelAndView 和 ModelMap 对象。

2.3.3 注解开发

2.3.2 节对 Spring MVC 的应用做了简单介绍，但所有处理器的编写最终都需要在配置文件中进行配置。然而，随着项目的扩大，整个项目中的处理器将会成倍增加，这将导致配置文件异常臃肿，虽可以按功能将其划分到不同的配置文件中并最终引入到主配置文件中，但不管如何操作，都逃不开处理器的配置。由于处理器太多且不方便维护，因此 Spring MVC 为开发人员提供了更为方便的处理器编写方式，即基于注解的方式。

前面已经介绍到处理器映射器主要将 request 对象映射到相应的处理器中，而处理器适配器主要负责处理器功能方法的调用，简言之，"一个管理 URL 地址，一个管理处理器的编写"。既然存在基于注解的处理器映射器和处理器适配器，就可以通过注解的方式来完成这两件事情。

（1）RequestMappingHandlerAdapter 处理器适配器规定需要作为处理器的 Bean 只要添加@Controller 注解即可。

（2）RequestMappingHandlerMapping 处理器映射器规定当需要完成处理器和 URL 地址的映射时，只要添加@RequestMapping 注解即可。

对这两个基于注解的处理器映射器和处理器适配器进行简单介绍之后，接下来改造 2.3.2 小节的应用。

1. 修改配置文件

```xml
<beans xmlns="http://www.springframework.org/schema/beans"
    xmlns:xsi="http://www.w3.org/2001/XMLSchema-instance"
    xmlns:mvc="http://www.springframework.org/schema/mvc"
    xmlns:context="http://www.springframework.org/schema/context"
    xsi:schemaLocation="http://www.springframework.org/schema/beans
    http://www.springframework.org/schema/beans/spring-beans-4.1.xsd
    http://www.springframework.org/schema/context
    http://www.springframework.org/schema/context/spring-context-4.1.xsd
    http://www.springframework.org/schema/mvc
    http://www.springframework.org/schema/mvc/spring-mvc-4.1.xsd">
    <context:component-scan base-package="com.woniu.springmvc.handler"/>

    <!-- 基于注解的处理器适配器和处理器映射器 -->
    <bean class="org.springframework.web.servlet.mvc.method.annotation.
RequestMappingHandlerMapping"></bean>
    <bean class="org.springframework.web.servlet.mvc.method.annotation.
RequestMappingHandlerAdapter"></bean>

    <!-- 视图解析器 -->
    <bean class="org.springframework.web.servlet.view.InternalResourceViewResolver">
        <!--指定视图文件的前缀 -->
        <property name="prefix" value="/WEB-INF/views/"/>
        <!-- 指定视图文件的后缀 -->
        <property name="suffix" value=".jsp"/>
    </bean>

</beans>
```

2. 修改处理器

```java
@Controller //标识该类为Spring MVC中的handler
@RequestMapping("/UserHandler")
public class UserHandler {

    @RequestMapping("/login") // 标识URL地址
    public String doLogin(HttpServletRequest request, HttpServletResponse response) {
        String userName = request.getParameter("userName");
        String password = request.getParameter("password");
        if ("admin".equals(password) && "admin".equals(userName)) {
```

```
            return "success";
        } else {
            return "fail";//逻辑视图名
        }
    }
}
```

在配置文件中需要扫描处理器所存在的包，使 Spring 的 IoC 容器得以识别。以上方式完成了处理器和 URL 地址的映射，现在的处理器不需要在配置文件中进行配置，只需添加相应注解即可。最后，改造配置文件，在 Spring MVC 中，通过 schema 为开发人员提供了 MVC 命名空间，以配置 Spring MVC 中相应的组件。以下方式即表示同时配置了基于注解的处理器适配器和处理器映射器。

V2-22　Spring MVC 注解开发

```
<!-- 注解驱动： -->
<mvc:annotation-driven/>
```

2.4　整合开发

通过前面的学习，相信读者应该掌握了使用 Spring 管理应用中的对象及对象间依赖关系的方法，以及使用 Spring MVC 处理 Web 请求和使用 MyBatis 完成数据持久化的方法，但企业开发应用中，往往需要几个框架联合使用。例如，Spring 提供了一致的事务管理以实现事务的统一，而恰好事务控制也是相当令人头疼的事情，何不将所有的事务都交由 Spring 处理，以使开发人员更加关注业务的实现呢？

2.4.1　搭建 Spring 开发环境

首先，对 Spring 的职责做简单介绍，Spring 主要提供对应用中对象的管理及事务的控制。这里沿用之前搭建的 Spring 环境，前面主要讲解的是通过 XML 的方式来完成对 Spring 的配置，为方便读者学习后面的章节，这里讲解在 Spring 3.0 以后引入的基于 JavaConfig 的方式搭建 Spring 应用的方法。

V2-23　SSM 整合 -1

1. 创建 Maven 项目并添加依赖

（1）创建项目。打开 Eclipse，创建 Maven 项目，具体可参照 2.1.2 小节的内容进行创建，但需要注意的是项目打包方式为"war"。因为是采用 JavaConfig 的方式来搭建项目，所以需要修改 pom.xml 文件，添加 Web 插件，代码如下。

```
<build>
    <plugins>
        <plugin>
            <groupId>org.apache.maven.plugins</groupId>
            <artifactId>maven-compiler-plugin</artifactId>
            <version>3.3</version>
            <configuration>
                <source>1.8</source>
                <target>1.8</target>
            </configuration>
        </plugin>
        <plugin>
            <groupId>org.apache.maven.plugins</groupId>
            <artifactId>maven-war-plugin</artifactId>
            <version>2.6</version>
            <configuration>
```

```xml
                <failOnMissingWebXml>false</failOnMissingWebXml>
            </configuration>
        </plugin>
    </plugins>
</build>
```

（2）添加依赖，代码如下。

```xml
<dependency>
    <groupId>org.springframework</groupId>
    <artifactId>spring-beans</artifactId>
    <version>4.3.12.RELEASE</version>
</dependency>
<dependency>
    <groupId>org.springframework</groupId>
    <artifactId>spring-core</artifactId>
    <version>4.3.12.RELEASE</version>
</dependency>
<dependency>
    <groupId>org.springframework</groupId>
    <artifactId>spring-context</artifactId>
    <version>4.3.12.RELEASE</version>
</dependency>
<dependency>
    <groupId>org.springframework</groupId>
    <artifactId>spring-aop</artifactId>
    <version>4.3.12.RELEASE</version>
</dependency>
<dependency>
    <groupId>org.springframework</groupId>
    <artifactId>spring-aspects</artifactId>
    <version>4.3.12.RELEASE</version>
</dependency>
<dependency>
    <groupId>org.springframework</groupId>
    <artifactId>spring-tx</artifactId>
    <version>4.3.12.RELEASE</version>
</dependency>
<dependency>
    <groupId>org.springframework</groupId>
    <artifactId>spring-jdbc</artifactId>
    <version>4.3.12.RELEASE</version>
</dependency>
<dependency>
    <groupId>mysql</groupId>
    <artifactId>mysql-connector-java</artifactId>
    <version>5.1.44</version>
</dependency>
```

2. 编写 Spring 的配置类 App

（1）分包。在项目中需要编写大量的 Java 类，所以在项目初期需要定义不同的包来管理类，以方便后期的维护。通常按照功能来分包，包结构如图 2-6 所示。

（2）在 config 包中创建 App.java 文件，代码如下。

```java
@Configuration
public class App {
}
```

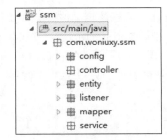

图 2-6　包结构

（3）配置数据源，代码如下。

```
@Bean
public DataSource dataSource(){
    DriverManagerDataSource dataSource=new DriverManagerDataSource();
    dataSource.setDriverClassName(driver);
    dataSource.setUrl(url);
    dataSource.setUsername(userName);
    dataSource.setPassword(password);
    return dataSource;
}
```

（4）引入外部的属性文件。在项目中，对于变化比较小的数据，一般将其提取到单独的属性文件中，再通过 Spring 的 IoC 容器注入进来。例如，将以上的 JDBC 的信息抽取到 jdbc.properties 中，再结合 Spring 提供的@PropertySource 注解和 PropertySourcesPlaceholderConfigurer 对象将属性文件中的数据直接注入到 App 类对应的属性上，具体代码如下。

V2-24　SSM 整合-2

```
@PropertySource("classpath:jdbc.properties")
public class App {

    @Value("${jdbc.driver}")
    private String driver;

    @Value("${jdbc.url}")
    private String url;

    @Value("${jdbc.username}")
    private String userName;

    @Value("${jdbc.password}")
    private String password;

    @Bean
    public static PropertySourcesPlaceholderConfigurer placeholderConfigurer(){
        return new PropertySourcesPlaceholderConfigurer();
    }
}
```

3．整合 Web 容器

（1）在 Spring 中提供 spring-web 模块来集成 Web 容器，在该模块中，Spring 提供了 Web 环境中的 WebApplicationContext 对象。修改 pom.xml，添加 Web 依赖，代码如下。

```xml
<dependency>
    <groupId>org.springframework</groupId>
    <artifactId>spring-web</artifactId>
    <version>4.3.12.RELEASE</version>
</dependency>
```

（2）编写监听器，用于监听 Servlet 容器的启动时间，初始化 Spring 容器。Web 包中提供了 WebApplicationInitializer 接口，该接口中的 onStartup 回调方法在 Web 容器启动完成之后会自动触发。在 config 包中创建 WebInitializer 类，用于实现 WebApplicationInitializer 接口，其内容如下。

```
public class WebInitializer implements WebApplicationInitializer {

    @Override
    public void onStartup(ServletContext servletContext) throws ServletException {
        AnnotationConfigWebApplicationContext act=new
AnnotationConfigWebApplicationContext();
        act.register(App.class);
        //和Servlet整合
        act.setServletContext(servletContext);
    }

}
```

2.4.2　Spring 集成 MyBatis

MyBatis 主要负责完成对数据的持久化，主要通过基于 Mapper 代理开发实现，而通过前面的学习可知，想要获取 Mapper 对象，需要先得到 SqlSession 对象，SqlSession 对象需要通过 SqlSessionFactory 获取，并且 SqlSessionFactory 对象是一个线程安全的对象，完全可以交由 Spring 进行管理。而 Spring 和 MyBatis 的集成主要就是 Spring 提供会话工厂的创建和事务的管理。

V2-25　SSM 整合-3

1. 添加 Spring 和 MyBatis 的整合依赖

```
<dependency>
        <groupId>org.mybatis</groupId>
        <artifactId>mybatis</artifactId>
        <version>3.4.4</version>
</dependency>
<dependency>
        <groupId>org.mybatis</groupId>
        <artifactId>mybatis-spring</artifactId>
        <version>1.3.1</version>
</dependency>
```

2. 添加会话工厂 Bean

在 src/main/resources 目录中创建 mappings 文件夹，用于存放 MyBatis 的所有映射文件，并注册到会话工厂中。

```
@Bean
    public SqlSessionFactory sqlSessionFactory() throws Exception{
        SqlSessionFactoryBean sqlSessionFactory=new SqlSessionFactoryBean();
        sqlSessionFactory.setDataSource(dataSource());
        //设置类型别名
sqlSessionFactory.setTypeAliasesPackage("com.woniuxy.ssm.entity");
        //注册映射文件
        sqlSessionFactory.setMapperLocations(new PathMatchingResourcePatternResolver().
getResources("classpath*:mappings/*.xml"));
        return sqlSessionFactory.getObject();
    }
```

3. 添加 Mapper 的扫描器

MyBatis 和 Spring 的整合包为用户提供了注解@MapperScan，使用它可以很方便地在基于 JavaConfig 的方式下完成对 Mapper 的扫描，但需要在 Mapper 接口中添加@Mapper 注解，修改 App 类，添加

@MapperScan 注解并指定需要扫描的包，代码如下。

```
@Configuration
@MapperScan(basePackages="com.woniuxy.ssm.mapper")
@PropertySource("classpath:jdbc.properties")
public class App {
```

4. 添加事务管理器

通过前面的学习，可以了解到 Spring 为开发人员提供了一致的事务管理器，不同的事务管理器支持不同的 ORM 框架，如 MyBatis 需要使用由 JDBC 包提供的基于数据源的事务管理器，代码如下。

```
@Bean
    public PlatformTransactionManager transactionManager(){
        DataSourceTransactionManager transactionManager=new DataSourceTransactionManager();
        transactionManager.setDataSource(dataSource());
        return transactionManager;
    }
```

5. 开启注解事务

在基于 JavaConfig 的环境下，只需要在配置类中添加@EnableTransactionManagement 注解即可开启注解事务。

```
@Configuration
@EnableTransactionManagement
@MapperScan(basePackages="com.woniuxy.ssm.mapper")
@PropertySource("classpath:jdbc.properties")
public class App {
}
```

V2-26 SSM 整合-4

2.4.3 Spring 集成 Spring MVC

通过前面的学习，读者可了解到 Spring MVC 主要用于处理 Web 请求，并且是 Spring 的 Web 模块，因此其可以和 Spring 无缝集成。

1. 添加 Spring MVC 的依赖

```
<dependency>
        <groupId>org.springframework</groupId>
        <artifactId>spring-webmvc</artifactId>
        <version>4.3.12.RELEASE</version>
</dependency>
```

2. 添加 Spring MVC 的配置类

在基于注解的 Spring MVC 配置中，只要配置视图解析器，并添加@EnableWebMvc 注解即可。同时，Spring MVC 也提供了 WebMvcConfigurerAdapter 类来定制配置，该类中提供了使开发人员能够容易地注册视图解析器、拦截器等的一系列方法。在 config 包中新建 WebMvcConfig 类，并继承 WebMvcConfigurerAdapter 类来完成对 Spring MVC 的配置，步骤如下。

（1）添加视图解析器。该实例中主要采用 JSP 作为视图引擎，所以配置 JstlView 来支持 JSTL 表达式和 JSP 视图。

```
@Configuration
public class WebMvcConfig extends WebMvcConfigurerAdapter{

    ViewResolver viewResolver(){
        InternalResourceViewResolver viewResolver=new InternalResourceViewResolver();
        viewResolver.setPrefix("/WEB-INF/views/");
        viewResolver.setSuffix(".jsp");
        viewResolver.setViewClass(JstlView.class);
        return viewResolver;
```

```
    }
    @Override
    public void configureViewResolvers(ViewResolverRegistry registry) {
        //注册视图解析器
        registry.viewResolver(viewResolver());
    }
```
（2）开启 WebMvc 支持。从前面的学习内容中了解到，Spring MVC 主要通过处理器来处理 Web 请求，在开启 Web 支持的同时指定了处理器所在的包。

```
@Configuration
@EnableWebMvc
@ComponentScan(basePackages="com.woniuxy.ssm.controller")
public class WebMvcConfig extends WebMvcConfigurerAdapter{
}
```

3. 注册前端控制器

在 2.3 节中，通过在 web.xml 文件中注册 Spring MVC 的中央控制器 DispatcherServlet 来拦截 Web 请求，那么基于 JavaConfig 的方式应如何实现呢？还记得 2.3 节中 Spring 是如何集成 Web 容器的吗？这里也采用相同的方式完成对中央控制器的注册，修改 WebInitializer 类，添加如下代码。

```
public class WebInitializer implements WebApplicationInitializer {
    @Override
    public void onStartup(ServletContext servletContext) throws ServletException {
        //配置Spring MVC的控制器
        Dynamic dynamic = servletContext.addServlet("SpringMVC", new DispatcherServlet(act));
        //设置后缀
        dynamic.addMapping("/");
        dynamic.setLoadOnStartup(1);
    }
}
```

4. 部署项目并启动 Web 服务器

将项目部署到 Web 容器中（本案例使用 Tomcat 8.0），启动服务器，如果出现以下提示信息，则说明整合完成，如图 2-7 所示。

```
10:12:33,351 DEBUG PropertySourcesPropertyResolver:92 - Could not find key 'spring.liveBeansView.mbeanDomain' in a
10:12:33,355 DEBUG DispatcherServlet:568 - Published WebApplicationContext of servlet 'springmvc' as ServletContex
10:12:33,355  INFO DispatcherServlet:508 - FrameworkServlet 'springmvc': initialization completed in 3964 ms
10:12:33,356 DEBUG DispatcherServlet:174 - Servlet 'springmvc' configured successfully
十一月 28, 2017 10:12:33 上午 org.apache.coyote.AbstractProtocol start
信息: Starting ProtocolHandler ["http-nio-8085"]
十一月 28, 2017 10:12:33 上午 org.apache.coyote.AbstractProtocol start
信息: Starting ProtocolHandler ["ajp-nio-8009"]
十一月 28, 2017 10:12:33 上午 org.apache.catalina.startup.Catalina start
信息: Server startup in 7645 ms
```

图 2-7 提示信息

2.5 开发实战

2.5.1 项目简介

此项目主要沿用 2.4 节中整合完的项目进行功能性演示，这里对 Web 系统中最典型的用户登录功能做简单介绍。

2.5.2 开发思路

鉴于在第 1 章中已经学习过 Web 开发的基础知识，也实现了简单的功能，因此这里不做过多介绍，只通过流程图的方式展示接下来将要做的事情。首先，由用户发送登录请求，交给中央控制器进行处理，再由处理器进行处理并根据处理的结果返回相应的视图，这里要实现的功能的流程如图 2-8 所示。

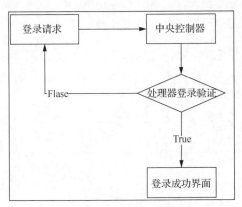

图 2-8 要实现的功能的流程

2.5.3 代码实现

1. 创建数据库表

```
create table t_user(
 id bigint(20) PRIMARY key auto_increment comment '编号',
 user_name varchar(60) not null comment '用户名称',
 cnname varchar(60) not null comment '姓名',
 sex tinyint(3) not null comment '性别',
 mobile varchar(20) not null comment '手机号码',
 email varchar(60) not null comment '电子邮件',
 note varchar(1024) comment '备注'
);
```

2. 添加测试数据

```
INSERT INTO 't_user' ('user_name', 'cnname', 'sex', 'mobile', 'email', 'note') VALUES
('zhangsan', '张三', '1', '12313', '123123', '蜗牛学院');
```

3. 编写 User 实体类

```java
public class User {

    private int id;
    private String user_name;
    private String cnname;
    private int sex;
    private String mobile;
    private String email;
    private String note;
}
```

4. 编写 UserMapper 接口

```java
@Mapper
public interface UserMapper {

    public User validateUserInfo(User user);
```

}

5. 编写 UserMapper 的映射文件

```xml
<?xml version="1.0" encoding="UTF-8" ?>
<!DOCTYPE mapper PUBLIC "-//mybatis.org//DTD Mapper 3.0//EN"
    "http://mybatis.org/dtd/mybatis-3-mapper.dtd">
<mapper namespace="com.woniuxy.ssm.mapper.UserMapper">
    <select id="validateUserInfo" parameterType="user" resultType="user">
        SELECT * from t_user
            where user_name=#{user_name} and password=#{password}
    </select>
</mapper>
```

6. 编写 UserService 类并添加用户校验方法

```java
@Service("userService")
@Transactional(readOnly=true)
public class UserService {

    @Resource
    private UserMapper userMapper;

    public User validateUserInfo(User user){
        return this.userMapper.validateUserInfo(user);
    }

}
```

7. 编写登录界面

```jsp
<%@ page language="java" contentType="text/html; charset=UTF-8"
    pageEncoding="UTF-8"%>
<!DOCTYPE html>
<html>
<head>
<meta http-equiv="Content-Type" content="text/html; charset=UTF-8">
<title>用户登录</title>
</head>
<body>
    <form action="" method="post">
        <table>
            <tr>
                <td>用户名：</td>
                <td>
                    <input type="text" name="user_name">
                </td>
            </tr>
            <tr>
                <td>密码：</td>
                <td>
                    <input type="password" name="password">
                </td>
            </tr>
            <tr>
                <td colspan="2" align="center">
                    <input type="submit" value="登录">
                    <input type="reset" value="重置">
                </td>
            </tr>
```

```
      </table>
    </form>
</body>
</html>
```

8. 编写登录成功界面

```
<%@ page language="java" contentType="text/html; charset=UTF-8"
    pageEncoding="UTF-8"%>
<!DOCTYPE html>
<html>
<head>
<meta http-equiv="Content-Type" content="text/html; charset=UTF-8">
<title>Insert title here</title>
</head>
<body>
    登录成功
</body>
</html>
```

9. 编写 UserController 类以处理用户请求

```
@Controller
public class LoginController {

    @Resource
    private UserService userService;

    @RequestMapping("/login")
    public String login(User user,HttpSession session){
        User userInfo = this.userService.validateUserInfo(user);
        if(userInfo!=null){
            session.setAttribute("user", userInfo);
            return "success";
        }else{
            return "login";
        }
    }
}
```

10. 部署启动 Web 容器

修改登录界面 form 表单的 action 属性为以上 LoginController 处理器的 URL 地址，部署启动 Tomcat 服务器，输入用户名和密码，检测登录逻辑是否正确。

本章中涉及的知识点比较多，先对目前企业开发项目中比较常用的开源框架做了简单介绍，使读者对其稍有概念，再通过 Spring 中提供的基于 JavaConfig 的方式对以上框架进行了整合应用，最后通过一个简单的实例验证了整合的效果。其中，对于 JavaConfig 的应用是本章的重点，也是后期学习 Spring Boot 的基础，希望读者重点把握。

V2-27　SSM 整合-5

第3章

Spring Boot

学习目标

（1）了解Spring Boot的自动装配。
（2）学会编写自定义starter。
（3）学会搭建应用框架。

本章导读

■通过第 2 章的学习，相信读者掌握了在项目中整合 Spring、Spring MVC 和 MyBatis 的方法，学会了在整合之后的基础架构上进行项目开发的过程。但在搭建环境的整个过程中，似乎编写真正和业务相关的代码（本身业务并不复杂）并没有花费多少时间，大部分时间花费在各种配置文件的编写上，而配置文件又是不可或缺的部分，且其中的内容在项目中大致是一致的。在项目开发过程中，讲求代码的复用，这些配置如何进行复用呢？Spring 为开发人员引入了 Spring Boot，并提供了大量的 starter，只需要在项目中添加相应的 starter 即可快速集成，节省了大量的开发时间。

3.1 Spring Boot 概述

3.1.1 了解 Spring Boot

随着动态语言的流行,Java 的开发显得格外笨重,需要编写大量的配置文件以整合第三方支持,虽然在 Spring 3.0 以后引入了基于 JavaConfig 的方式来配置 Spring,但是这些工作没有任何难度却占用了大量的开发时间,还容易出错。

基于以上情况,Spring Boot 应运而生,其以"习惯优于配置"的理念使用户快速搭建开发环境。Spring Boot 使得开发人员可以很容易地创建独立的、准生产级别的 Spring 应用,并且只需要通过执行简单的 Java-jar 命令即可运行整个项目。

以下是 Spring Boot 为开发人员承诺的目标。
(1) 为所有 Spring 开发人员提供更快、更广泛的入门体验,开箱即用。
(2) 提供大型项目(如嵌入式服务器、安全性、指标、运行状况检查、外部配置)通用的一系列非功能性的功能。
(3) 绝对不会有代码生成,也不需要 XML 配置。

V3-1 Spring Boot 简介

3.1.2 Spring Boot 的核心功能

(1) 独立运行的 Spring 项目。Spring Boot 可以以 JAR 包的方式独立运行整个项目,并支持内嵌数据库和 Servlet 容器,其内嵌的 Servlet 容器支持 Tomcat、Jetty 和 Undertow,开发人员可以根据实际需求方便地进行切换。

(2) 其所提供的 starter 可简化 Maven 配置。通过前面的学习,可知 Spring 提供了 XML 和 JavaConfig 方式来整合第三方框架,但配置文件随着框架的不同存在相应的差异,且容易出错。出于这一原因,Spring Boot 为开发人员提供了大量的 starter 来简化配置。常用的 starter 如表 3-1 所示。

表 3-1 Spring Boot 常用的 starter

名称	说明
spring-boot-starter	核心组件,提供对自动配置、日志及 YAML 的支持
spring-boot-starter-activemq	用于集成 Apache 的基于 JMS 协议的 ActiveMQ 消息通信框架
spring-boot-starter-amqp	用于集成基于 AMQP 的消息通信 Rabbit MQ
spring-boot-starter-aop	用于支持 Spring AOP 和 AspectJ,实现面向切面的编程
spring-boot-starter-artemis	用于支持 Apache Artemis 的 JMS 消息服务
spring-boot-starter-batch	用于支持 Spring 的批处理服务
spring-boot-starter-cache	用于 Spring 对缓存的支持
spring-boot-starter-data-elasticsearch	用于整合 ElasticSearch 搜索和分析引擎的支持
spring-boot-starter-data-mongodb	用于整合文档型数据库 mongodb 的支持
spring-boot-starter-data-neo4j	用于整合图形数据库 Neo4j 的支持
spring-boot-starter-data-rest	用于整合开发 REST 应用
spring-boot-starter-data-solr	用于整合全文检索引擎 Solr
spring-boot-starter-data-jpa	用于集成 Hibernate 的 JPA,实现数据的持久化

续表

名称	说明
spring-boot-starter-data-redis	用于集成非关系型数据库 Redis
spring-boot-starter-freemarker	用于集成模板引擎 FreeMarker
spring-boot-starter-groovy-templates	用于集成模板引擎 Groovy
spring-boot-starter-jdbc	用于集成 Apache 提供的数据库连接池
spring-boot-starter-mail	用于整合 JMail 简化邮件发送
spring-boot-starter-security	用于整合 Spring 的安全性框架 Spring Security
spring-boot-starter-test	用于整合 Spring 的单元测试框架
spring-boot-starter-web	用于集成 Spring MVC
spring-boot-starter-actuator	用于提供应用监控

以上只罗列了官网提供的部分 starter，有兴趣的读者可以查阅官方文档查看完整的 starter，同时，除了官方提供的 starter 以外还有一些第三方的 starter 可以用于集成特定的第三方库。正因为各种 starter 的存在，加上 Spring 的自动装配，两者的结合简化了整个整合过程，因此掌握 starter 的编写也是日常项目开发中的重要环节，这点将在 3.2 节中重点进行讲解。

（3）自动配置 Spring。
（4）准生产环境的应用监控。
（5）无代码生成和 XML 配置。

3.1.3　Spring Boot 示例

下面来了解 Spring Boot 的快速搭建企业级开发应用功能的使用过程。

1．创建项目

通过前面的介绍，可以了解到 Spring Boot 可以快速搭建环境并以执行 Java –jar 命令的方式运行整个项目，这意味着项目需要以 JAR 的方式进行打包。然而，开发者需要的却是一个 Web 项目。Spring Boot 支持内嵌的 Servlet 容器，这正是其强大之处，可以通过一个普通的 JAR 包运行 Web 应用，这一特点也是目前市面上采用 Spring Boot 来构建微服务的主要原因，同时，Spring Boot 为开发人员提供了传统的 WAR 方式来部署 Web 应用。这里先创建 Spring Boot 项目，如图 3-1 所示。

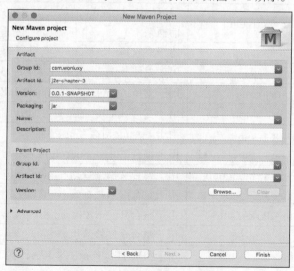

图 3-1　创建 Spring Boot 项目

2. 添加 Spring Boot 的继承

创建好项目之后，需要通过 Maven 中提供的继承方式将当前项目继承依赖于 Spring Boot，即修改 pom.xml 文件，添加如下内容。

```xml
<parent>
    <groupId>org.springframework.boot</groupId>
    <artifactId>spring-boot-starter-parent</artifactId>
    <version>1.4.7.RELEASE</version>
</parent>
```

3. 集成 Spring MVC

Spring Boot 为开发人员提供了大量的 starter 来快速集成第三方框架，如需要集成 Spring MVC 时，只要添加其 starter 依赖即可，代码如下。

```xml
<dependencies>
    <dependency>
        <groupId>org.springframework.boot</groupId>
        <artifactId>spring-boot-starter-web</artifactId>
    </dependency>
</dependencies>
```

引入该 starter 之后，表示项目中已经集成了 Spring MVC 的支持，同时可以看到项目中是否添加了相应的 JAR 文件，展开项目中的"maven dependencies"节点，可以看到熟悉的 JAR 文件，如图 3-2 中框选部分所示。

这两个文件一个文件用于整合 Servlet 容器，另一个文件是 Spring MVC 的核心依赖文件，这两个文件在前面章节的学习中已经提到过，这里不再赘述。说到 Servlet 容器，在引入的依赖中还可以看到引入的内嵌 Servlet 容器 Tomcat 的依赖（Spring Boot 除了支持 Tomcat 以外，还支持 Jetty 和 Undertow，默认采用 Tomcat，可以根据自己的需求切换使用的 Servlet 容器），如图 3-3 所示。

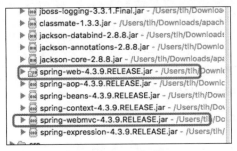

图 3-2　JAR 文件

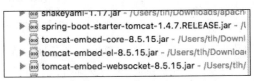

图 3-3　Servlet 容器

4. 添加插件

修改 pom.xml 文件，添加可以将项目打包为可执行 JAR 文件的 Spring Boot 插件，在开发阶段可以不添加该插件，但在项目进行打包时必须添加该插件，否则在执行的时候会因找不到主类而导致项目无法正常启动。

```xml
<build>
    <plugins>
        <plugin>
            <groupId>org.apache.maven.plugins</groupId>
            <artifactId>maven-compiler-plugin</artifactId>
            <configuration>
                <source>1.8</source>
                <target>1.8</target>
            </configuration>
        </plugin>
```

```xml
            <!--Spring Boot打包插件 -->
            <plugin>
                <groupId>org.springframework.boot</groupId>
                <artifactId>spring-boot-maven-plugin</artifactId>
            </plugin>
        </plugins>
    </build>
```

5. 编写配置文件

虽然 Spring Boot 中不需要编写过多的 XML 文件，但是关于核心的配置仍需要编写在配置文件中，如 Servlet 容器的端口号、数据库信息等。Spring Boot 为开发人员提供了 properties 和 yml 两种文件格式的配置文件，不管采用哪种格式，其默认使用的配置文件的名称都为 application。在 resources 目录中创建 application.properties 文件，添加如下内容。

```
server.port=8085
```

Spring Boot 中提供了很多属性，以供开发人员修改默认配置，如这里将 Tomcat 的默认端口号修改为 8085。

6. 编写启动类

在编写启动类之前应对整个 Spring Boot 项目的包结构有些概念，以免项目在启动过程中出现不必要的错误，这里 Spring Boot 的项目结构如图 3-4 所示。

```
com
+- example
    +- myproject
        +- Application.java
        |
        +- domain
        |   +- Customer.java
        |   +- CustomerRepository.java
        |
        +- service
        |   +- CustomerService.java
        |
        +- web
            +- CustomerController.java
```

图 3-4　Spring Boot 的项目结构

其中，Application.java 是整个 Spring Boot 的启动类，其内容也非常简单，和普通 Java 应用的主类没有任何区别，只要提供 main 方法即可，其内容如下。

```java
package com.woniuxy.springboot;

import org.springframework.boot.SpringApplication;
import org.springframework.boot.autoconfigure.SpringBootApplication;

@SpringBootApplication
public class Application {

    public static void main(String[] args) {
        SpringApplication.run(Application.class, args);
    }
}
```

上面在 com.woniuxy.springboot 包中创建了 Application 主类作为整个应用的启动类，对该类中涉及的要点做以下说明。

（1）@SpringBootApplication：该注解是一个组合注解，其组合了@SpringBootConfiguration、@EnableAutoConfiguration 和@ComponentScan 3 个注解，完成自动装配及自动扫描功能。其中，

@ComponentScan 会默认扫描和主类同级的包中添加了 service、controller、repository 和 component 注解的对象并交由 Spring 进行管理，这正因此，主类最好处于最顶层。

（2）在 main 方法中通过 SpringApplication 的 run 方法来启动整个应用。

7. 编写处理器

新建 controller 包，在其下创建 HelloController 类。

```
@RestController
public class HelloController {
    @RequestMapping("/")
    public String hello(){
        return "hello,spring boot!!";
    }
}
```

其中，@RestController 注解为一个组合注解，组合了@Controller 和@Responsebody 注解，用于将返回的结果放入 Response Body。

8. 启动并访问项目

启动 Application 主类，若看到如图 3-5 所示的信息即表示应用启动成功。

图 3-5 Spring Boot 启动成功

通过启动日志可以看到，Tomcat 使用的是之前配置的 8085 端口，此时打开浏览器，在地址栏中输入 "http://localhost:8085" 即可访问 Spring Boot 项目，如图 3-6 所示。

图 3-6 访问 Spring Boot 项目

至此，已经完成通过 Spring Boot 来搭建 Spring MVC 的环境，在整个应用中没有编写任何有关 Spring MVC 的配置代码，只是简单地在项目中引入了 starter 依赖。这就是 Spring Boot 的神奇之处，不需要编写任何配置文件，将更多的精力放到具体的业务实现上。

V3-2 Spring Boot 示例

3.2 Spring Boot 核心

3.2.1 自动配置

通过 3.1 节的例子，相信读者应该感受到了 Spring Boot 的强大之处，引入 Web 的 starter 就可以在项目中快速地整合开发 Spring MVC 应用，那么这是如何实现的呢？

首先，查看项目中添加的依赖，从引入的 JAR 包中可发现一个名为 spring-boot-autoconfigure-x.x.jar 的 JAR 包，其中存放了 Spring Boot 关于自动配置的所有源码，从展开的包名中可以清晰地看到不同框架的整合实现，如图 3-7 所示。

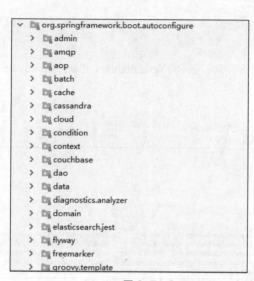

图 3-7 Spring Boot 自动配置

由图 3-7 可知 Spring Boot 会根据不同的环境完成自动配置。3.1 节的示例在主类中添加了 @SpringBootApplication 注解，它是一个组合注解，组合了自动配置和自动扫描功能，现在来看该注解的定义。

```
@Target(ElementType.TYPE)
@Retention(RetentionPolicy.RUNTIME)
@Documented
@Inherited
@SpringBootConfiguration
@EnableAutoConfiguration
@ComponentScan(excludeFilters = @Filter(type = FilterType.CUSTOM, classes = TypeExcludeFilter.class))
public @interface SpringBootApplication {
}
```

关于元注解这里不做过多介绍，主要看 EnableAutoConfiguration 注解。从字面上可以看出其表示"开启自动配置"，那么这个注解到底有什么作用呢？下面先看其声明。

```
@Target(ElementType.TYPE)
@Retention(RetentionPolicy.RUNTIME)
@Documented
@Inherited
@AutoConfigurationPackage
@Import(EnableAutoConfigurationImportSelector.class)
public @interface EnableAutoConfiguration {

    String ENABLED_OVERRIDE_PROPERTY = "spring.boot.enableautoconfiguration";
```

```java
/**
 * Exclude specific auto-configuration classes such that they will never be applied.
 * @return the classes to exclude
 */
Class<?>[] exclude() default {};

/**
 * Exclude specific auto-configuration class names such that they will never be
 * applied.
 * @return the class names to exclude
 * @since 1.3.0
 */
String[] excludeName() default {};
}
```

从其声明中并没有发现太有价值的信息，但是可注意到，它通过 Import 注解引入了一个 Selector，下面来看该类的实现。

```java
public class EnableAutoConfigurationImportSelector
    implements DeferredImportSelector, BeanClassLoaderAware, ResourceLoaderAware,
    BeanFactoryAware, EnvironmentAware, Ordered {
    @Override
    public String[] selectImports(AnnotationMetadata metadata) {
        if (!isEnabled(metadata)) {
            return NO_IMPORTS;
        }
        try {
            AnnotationAttributes attributes = getAttributes(metadata);
            List<String> configurations = getCandidateConfigurations(metadata,
                attributes);
            configurations = removeDuplicates(configurations);
            Set<String> exclusions = getExclusions(metadata, attributes);
            configurations.removeAll(exclusions);
            configurations = sort(configurations);
            recordWithConditionEvaluationReport(configurations, exclusions);
            return configurations.toArray(new String[configurations.size()]);
        }
        catch (IOException ex) {
            throw new IllegalStateException(ex);
        }
    }
}
```

上面截取的是 Selector 中的关键实现，发现其调用了一个方法名为 getCandidateConfigurations 的方法，此方法的实现如下。

```java
protected List<String> getCandidateConfigurations(AnnotationMetadata metadata,
        AnnotationAttributes attributes) {
    List<String> configurations = SpringFactoriesLoader.loadFactoryNames(
            getSpringFactoriesLoaderFactoryClass(), getBeanClassLoader());
    Assert.notEmpty(configurations,
            "No auto configuration classes found in META-INF/spring.factories. If you "
            + "are using a custom packaging, make sure that file is correct.");
    return configurations;
}
```

从此方法的实现及抛出的异常可以看出，Spring Boot 通过 SpringFactoriesLoader 的 loadFactoryNames 方法读取了一个名为 spring.factories 的文件，而 autoconfigure 的 JAR 包中就有这个文件，其内容如下。

```
# Initializers
org.springframework.context.ApplicationContextInitializer=\
org.springframework.boot.autoconfigure.SharedMetadataReaderFactoryContextInitializer,\
org.springframework.boot.autoconfigure.logging.AutoConfigurationReportLoggingInitializer

# Application Listeners
org.springframework.context.ApplicationListener=\
org.springframework.boot.autoconfigure.BackgroundPreinitializer

# Auto Configure
org.springframework.boot.autoconfigure.EnableAutoConfiguration=\
org.springframework.boot.autoconfigure.admin.SpringApplicationAdminJmxAutoConfiguration,\
org.springframework.boot.autoconfigure.aop.AopAutoConfiguration,\
org.springframework.boot.autoconfigure.amqp.RabbitAutoConfiguration,\
org.springframework.boot.autoconfigure.MessageSourceAutoConfiguration,\
org.springframework.boot.autoconfigure.PropertyPlaceholderAutoConfiguration,\
org.springframework.boot.autoconfigure.batch.BatchAutoConfiguration,\
org.springframework.boot.autoconfigure.cache.CacheAutoConfiguration,\
org.springframework.boot.autoconfigure.cassandra.CassandraAutoConfiguration,\
org.springframework.boot.autoconfigure.cloud.CloudAutoConfiguration,\
org.springframework.boot.autoconfigure.context.ConfigurationPropertiesAutoConfiguration,\
org.springframework.boot.autoconfigure.couchbase.CouchbaseAutoConfiguration,\
```

此文件中存储了所有自动配置的声明，随便打开其中一个配置的声明，其内容如下。

```
@Configuration
@ConditionalOnWebApplication
@ConditionalOnClass({ Servlet.class, DispatcherServlet.class,
        WebMvcConfigurerAdapter.class })
@ConditionalOnMissingBean(WebMvcConfigurationSupport.class)
@AutoConfigureOrder(Ordered.HIGHEST_PRECEDENCE + 10)
@AutoConfigureAfter(DispatcherServletAutoConfiguration.class)
public class WebMvcAutoConfiguration {

    public static String DEFAULT_PREFIX = "";

    public static String DEFAULT_SUFFIX = "";
```

以上是完成 Spring MVC 自动配置的核心声明，正如前面讲的，Spring Boot 中的所有配置都是基于 JavaConfig 来实现的，细心的读者会发现其声明上多了一些形如 ConditionalOnXXX 的注解，这些注解具有什么作用呢？这些是 Spring 提供的条件注解，结合不同的条件进行配置，正因这些条件注解的存在，Spring Boot 才得以在不同的环境下完成相应的自动配置。其内置的核心的条件注解如下。

（1）@ConditionalOnClass：检查类加载器中是否存在对应的类，如果存在，则被注解修饰的类就有资格被 Spring 容器所注册，否则会被跳过。

（2）@ConditionalOnBean：表示在容器中指定的 Bean 存在的条件下，被注解修饰的类就有资格被 Spring 容器所注册，否则会被跳过。

（3）@ConditionalOnExpression：以 SpEL 表达式作为判断条件。

（4）@ConditionalOnJava：以 JVM 版本作为判断条件。

（5）@ConditionalOnJndi：在 JNDI 存在的条件下查找指定的位置。

（6）@ConditionalOnMissingBean：在容器中没有指定 Bean 的情况下，被注解修饰的类就有资格被 Spring 容器所注册，否则会被跳过。

（7）@ConditionalOnMissingClass：在类路径下没有指定的类的条件下，被注解修饰的类就有资格被 Spring 容器所注册，否则会被跳过。

（8）@ConditionalOnNotWebApplication：在前项目不是 Web 项目的条件下，被注解修饰的类就有资格被 Spring 容器所注册，否则会被跳过。

（9）@ConditionalOnProperty：指定的属性是否有指定的值，如果有，则被注解修饰的类就有资格被 Spring 容器所注册，否则会被跳过。

（10）@ConditionalOnResource：类路径是否有指定的值，如果有，则被注解修饰的类就有资格被 Spring 容器所注册，否则会被跳过。

（11）@ConditionalOnSingleCandidate：在指定 Bean 在容器中只有一个实例，或者虽然有多个实例但是指定首选的 Bean 的情况下。

（12）@ConditionalOnWebApplication：在当前项目是 Web 项目的条件下。

这些条件注解都组合了 @Conditional 元注解，只是使用了不同的条件。下面来简单分析 @ConditionalOnWebApplication 注解。

```
@Target({ ElementType.TYPE, ElementType.METHOD })
@Retention(RetentionPolicy.RUNTIME)
@Documented
@Conditional(OnWebApplicationCondition.class)
public @interface ConditionalOnWebApplication {
}
```

从其源码可以看出，此注解使用的条件是 OnWebApplicationCondition。下面来看这个条件是如何构造的。

```
@Order(Ordered.HIGHEST_PRECEDENCE + 20)
class OnWebApplicationCondition extends SpringBootCondition {

    private static final String WEB_CONTEXT_CLASS = "org.springframework.web.context."
        + "support.GenericWebApplicationContext";

    @Override
    public ConditionOutcome getMatchOutcome(ConditionContext context,
            AnnotatedTypeMetadata metadata) {
        boolean required = metadata
            .isAnnotated(ConditionalOnWebApplication.class.getName());
        ConditionOutcome outcome = isWebApplication(context, metadata, required);
        if (required && !outcome.isMatch()) {
            return ConditionOutcome.noMatch(outcome.getConditionMessage());
        }
        if (!required && outcome.isMatch()) {
            return ConditionOutcome.noMatch(outcome.getConditionMessage());
        }
        return ConditionOutcome.match(outcome.getConditionMessage());
    }

    private ConditionOutcome isWebApplication(ConditionContext context,
            AnnotatedTypeMetadata metadata, boolean required) {
        ConditionMessage.Builder message = ConditionMessage.forCondition(
                ConditionalOnWebApplication.class, required ? "(required)" : "");
        if (!ClassUtils.isPresent(WEB_CONTEXT_CLASS, context.getClassLoader())) {
            return ConditionOutcome
                .noMatch(message.didNotFind("web application classes").atAll());
        }
        if (context.getBeanFactory() != null) {
            String[] scopes = context.getBeanFactory().getRegisteredScopeNames();
            if (ObjectUtils.containsElement(scopes, "session")) {
                return ConditionOutcome.match(message.foundExactly("'session' scope"));
            }
```

```
        }
        if (context.getEnvironment() instanceof StandardServletEnvironment) {
            return ConditionOutcome
                    .match(message.foundExactly("StandardServletEnvironment"));
        }
        if (context.getResourceLoader() instanceof WebApplicationContext) {
            return ConditionOutcome.match(message.foundExactly("WebApplicationContext"));
        }
        return ConditionOutcome.noMatch(message.because("not a web application"));
    }
}
```

从 isWebApplication 方法可以看出，判断条件如下。

（1）GenericWebApplicationContext 的 class 文件是否在类路径中存在。

（2）容器中是否有名为 session 的 scope 作用域。

（3）当前环境是否为 StandardServletEnvironment。

（4）当前的 ResourceLoader 是否为 WebApplicationContext。

通过前面的讲解，读者应对 Spring Boot 的自动装配有了大致的体会，即通过 JavaConfig 的方式进行配置，再结合不同的 Conditional 注解实现不同环境下的自动装配。注意需要在 spring.factories 文件中进行注册，以使 Spring Boot 扫描到该自动配置类。

V3-3　Spring Boot 自动装配原理

3.2.2　自定义 starter

3.2.1 小节对 Spring Boot 自动装配的原理做了详细说明，下面来实现一个自定义的 starter。对 3.1 节的实例程序中显示的数据进行处理，提取一个 HelloService 服务类，以实现对该服务类的自动装配。

（1）创建项目，图 3-8 所示为创建的 Maven 项目。

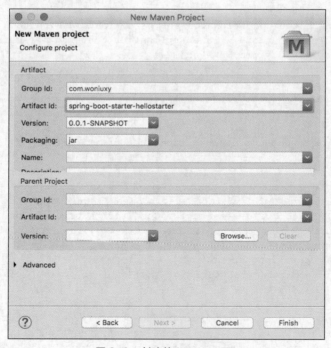

图 3-8　创建的 Maven 项目

（2）添加依赖，代码如下。

```xml
<dependencies>
    <dependency>
        <groupId>org.springframework.boot</groupId>
        <artifactId>spring-boot-autoconfigure</artifactId>
        <version>1.4.7.RELEASE</version>
    </dependency>
</dependencies>
```

（3）编写 HelloService 类，代码如下。

```java
public class HelloService {

    private String message;

    public String getMessage() {
        return message;
    }

    public void setMessage(String message) {
        this.message = message;
    }

}
```

（4）编写配置属性注入类 HelloServiceProperties，代码如下。

```java
@ConfigurationProperties(prefix="woniuxy.hello")
public class HelloServiceProperties {

    private String message;

    public String getMessage() {
        return message;
    }

    public void setMessage(String message) {
        this.message = message;
    }

}
```

（5）编写自动配置类 HelloServiceAutoConfiguration，代码如下。

```java
@Configuration
//开启属性配置
@EnableConfigurationProperties(HelloServiceProperties.class)
//只有在类路径中存在HelloService时才进行自动配置
@ConditionalOnClass(HelloService.class)
//总开关
@ConditionalOnProperty(prefix = "woniuxy.hello", value = "enabled", matchIfMissing = true)
public class HelloServiceAutoConfiguration {

    @Autowired
    private HelloServiceProperties helloServiceProperties;

    @Bean
    @ConditionalOnMissingBean(HelloService.class)
    //若Bean不存在，则创建Bean
```

```java
    public HelloService helloService() {
        HelloService helloService = new HelloService();
        helloService.setMessage(helloServiceProperties.getMessage());
        return helloService;
    }
}
```

（6）在 resources 目录中创建 META-INF/spring.factories 文件，代码如下。

```
org.springframework.boot.autoconfigure.EnableAutoConfiguration=\
com.woniuxy.springboot.service.config.HelloServiceAutoConfiguration
```

（7）修改 3.1 节中的 pom.xml 文件并引入该定义的 starter，代码如下。

```xml
<dependency>
    <groupId>com.woniuxy</groupId>
    <artifactId>spring-boot-starter-hello</artifactId>
    <version>0.0.1-SNAPSHOT</version>
</dependency>
```

（8）在 application.properties 文件中添加如下内容。

```
woniuxy.hello.message=hello,woniuxy
```

（9）在 hellocontroller 中引入自定义的 helloservice，代码如下。

```java
@RestController
public class HelloController {

    @Resource
    private HelloService helloService;

    @RequestMapping("/")
    public String hello(){
        return this.helloService.getMessage();
    }
}
```

（10）启动项目并在浏览器中访问 Spring Boot 项目，如图 3-9 所示。

图 3-9　在浏览器中访问 Spring Boot 项目

通过浏览器的输出可以看到显示的是在 application.properties 文件中编写的信息，如此即实现了在项目中对 HelloService 的自动装配。

本章讲解了如何使用 Spring Boot 来快速搭建开发环境，并讲解了其自动配置实现的原理，最后结合自定义的 starter 讲解了其运作的机制，相信读者通过本章的学习会有所收获。

第4章

Spring Data

本章导读

■在第2章中讲解了如何使用MyBatis完成数据的持久化，但其所针对的是关系型数据库。在实际项目开发中，随着项目规模的扩大，需要引入非关系型数据库以提高系统性能（如采用Redis处理缓存以提升数据库的性能），这意味着在项目中需要针对这种非关系型数据库编写一套存储实现，这无形中增加了开发时间。那么是否存在一种可针对所有平台进行数据存储的方式呢？Spring提出了Spring Data来统一关系型和非关系型数据存储，为上层实现数据的存储。

学习目标

（1）了解Spring Data。
（2）掌握使用Spring Data JPA实现关系型数据存储的操作。
（3）掌握使用Spring Data Redis实现对非关系型数据库Redis的操作。

4.1 数据持久化

4.1.1 了解数据持久化

在应用中，计算及其产生的数据都存储在内存中，如果不对其进行存储，则其在应用退出之后会丢失。所谓的数据持久化就是将内存中的数据模型转换为存储模型，将这些数据固化到某种介质中，从而保证这些数据在应用下一次开启的时候也能使用。

4.1.2 常用的数据持久化技术

前面对数据持久化的概念做了简单介绍，简而言之，数据持久化就是将内存中的数据固化到磁盘中的过程。数据持久化是一个过程，而在这个过程中需要借助的具体实现方式就称为数据持久化技术，如在 Java 平台中，Sun 公司提供的 Java 数据库连接（Java Database Connectivity，JDBC）规范可实现对关系型数据的存储；通过 Jedis 可以实现对 Redis 的操作。不同的平台采用的方式是不一样的，下面针对几种常用的数据持久化技术进行说明。

1. JDBC

JDBC 是一种用于执行 SQL 语句的 Java API，可以为多种关系型数据库提供统一访问，其由一组使用 Java 语言编写的类和接口组成。

2. Java 持久层 API

Java 持久层 API（Java Persistence API，JPA）是使用注解或 XML 来描述对象-关系表的映射关系，其将运行期的实体对象持久化到数据库中。

3. Jedis

Jedis 就是集成了 Redis 的一些命令操作，并封装了 Redis 的 Java 客户端，其提供了连接池管理，为上层应用提供了操作 Redis 更为便捷的方式。

4.2 持久化实现

4.1 节对数据持久化及数据持久化技术做了简单介绍，本节将列举针对关系型数据库和非关系型数据库的具体实现方法。其中，非关系型数据库种类较多，并且实现方式不尽相同，这里主要针对 Redis 进行简单介绍。后面的章节会对 Redis 的应用做更为深入的介绍，想提前了解这部分内容的读者可以阅读本书第 7 章。

4.2.1 关系型数据库的持久化实现

前面在介绍常用的数据持久化技术时提到的 JDBC 和 JPA 都是针对关系型数据库的技术，对于 JDBC 的实现方式这里不做过多介绍，这是 Java 开发人员实现数据持久化应掌握的最基本的技术，也是很多上层框架最底层的实现，例如，Hibernate ORM 和 MyBatis 都是对 JDBC 的包装。

JPA 实质上是一个规范 Java 的接口，其实现商有 Hibernate、OpenJPA、TopLink 等。这里的 Hibernate 指的是其对 JPA 的实现，和 Hibernate ORM 没有任何关系，为了承接后面要讲解的 Spring Data JPA，这里主要对 JPA 的应用做简单介绍。

1. 实验准备

打开 MySQL 客户端，执行以下命令，创建用户表。

```
create database jpa;
use jpa;
```

```sql
create table users(
 id int primary key auto_increment,
 name varchar(100),
 password varchar(32),
 address varchar(100),
 birthday timestamp
);
```

2. 创建项目并添加依赖

通过 Eclipse 创建 Maven 的 Java 项目，并修改 pom.xml 文件，添加如下依赖。

```xml
<dependencies>
    <dependency>
        <groupId>mysql</groupId>
        <artifactId>mysql-connector-java</artifactId>
        <version>5.1.44</version>
    </dependency>
    <dependency>
        <groupId>org.hibernate</groupId>
        <artifactId>hibernate-entitymanager</artifactId>
        <version>5.2.12.Final</version>
    </dependency>
    <dependency>
        <groupId>log4j</groupId>
        <artifactId>log4j</artifactId>
        <version>1.2.17</version>
    </dependency>
    <dependency>
        <groupId>org.slf4j</groupId>
        <artifactId>slf4j-log4j12</artifactId>
        <version>1.7.21</version>
    </dependency>
    <dependency>
        <groupId>junit</groupId>
        <artifactId>junit</artifactId>
        <version>4.12</version>
        <scope>test</scope>
    </dependency>
</dependencies>
```

3. 编写配置文件

在项目根目录的 resources 目录中创建 META-INF 文件夹，添加 persistence.xml 文件，并为其添加如下信息。

```xml
<?xml version="1.0" encoding="UTF-8"?>
<persistence version="2.0"
    xmlns="http://java.sun.com/xml/ns/persistence"
    xmlns:xsi="http://www.w3.org/2001/XMLSchema-instance"
    xsi:schemaLocation="http://java.sun.com/xml/ns/persistence
    http://java.sun.com/xml/ns/persistence/persistence_2_0.xsd">

    <persistence-unit name="jpa" transaction-type="RESOURCE_LOCAL">
        <!-- 配置数据库信息 -->
        <properties>
            <property name="hibernate.connection.driver_class" value="com.mysql.jdbc.Driver" />
            <property name="hibernate.connection.url" value="jdbc:mysql://localhost:3306/jpa" />
```

```xml
            <property name="hibernate.connection.username" value="数据库用户名" />
            <property name="hibernate.connection.password" value="数据库密码" />

            <property name="hibernate.dialect" value="org.hibernate.dialect.MySQL5Dialect" />

            <property name="hibernate.show_sql" value="true" />
            <property name="hibernate.hbm2ddl.auto" value="update" />
        </properties>

    </persistence-unit>

</persistence>
```

4. 编写实体类

```java
public class User {

    private Integer id;
    private String name;
    private String password;
    private String address;
    private Date birthday;
    //getter\setter
}
```

5. 编写映射文件

```java
@Entity
@Table(name = "users")
public class User {

    @Id
    @GeneratedValue(strategy = GenerationType.IDENTITY)
    private Integer id;
    private String name;
    private String password;
    private String address;
    private Date birthday;
```

在 JPA 中，通过在实体类中添加相应的注解来完成实体类和数据库表之间的映射关系。JPA 中常用的注解如下。

（1）Entity：标识该类为 JPA 中的实体类对象。

（2）Table：用于实体类和数据库表之间的映射，其中，name 属性用于设置映射到数据库的表名。

（3）Id：用于标识实体类中的主键，该注解必须添加，否则 JPA 中提供的部分 API 将无法使用。

（4）GeneratedValue：设置主键的生成策略，这里设置为自增。

（5）GenericGenerator：用于定义主键生成策略。

（6）Column：用于实体类属性和数据库表列的映射，其中，name 属性用于设置映射到数据库表中的列名。该注解可以不添加，此时将使用属性名作为数据库表列名的一一对应映射。

（7）Transient：用于处理非持久化的属性，标识该属性的数据在执行持久化操作时不会保存到数据库中。

6. 在配置文件中添加实体信息

```xml
<class>com.woniuxy.j2e.entity.User</class>
```

7. 测试

在项目根目录的 src/test/java 文件夹中创建 UserTest 类进行测试，实现对 User 表的添加、删除、更新和查询操作。

（1）对 User 表的添加操作，代码如下。

```java
@Test
public void save(){
    EntityManagerFactory entityManagerFactory = Persistence.createEntityManagerFactory("jpa");
    EntityManager entityManager = entityManagerFactory.createEntityManager();
    entityManager.getTransaction().begin();

    User user=new User();
    user.setAddress("成都");
    user.setBirthday(new Date());
    user.setName("admin");
    user.setPassword("admin");

    entityManager.persist(user);

    entityManager.getTransaction().commit();

    entityManager.close();
    entityManagerFactory.close();
}
```

（2）对 User 表的删除操作，代码如下。

```java
@Test
public void delete(){
    EntityManagerFactory entityManagerFactory = Persistence.createEntityManagerFactory("jpa");
    EntityManager entityManager = entityManagerFactory.createEntityManager();
    entityManager.getTransaction().begin();

    User user = entityManager.find(User.class, 1);
    entityManager.remove(user);

    entityManager.getTransaction().commit();

    entityManager.close();
    entityManagerFactory.close();
}
```

（3）对 User 表的更新操作，代码如下。

```java
@Test
public void update(){
    EntityManagerFactory entityManagerFactory = Persistence.createEntityManagerFactory("jpa");
    EntityManager entityManager = entityManagerFactory.createEntityManager();
    entityManager.getTransaction().begin();

    User user=new User();
    user.setId(2);
    user.setAddress("北京");

    entityManager.merge(user);

    entityManager.getTransaction().commit();

    entityManager.close();
```

```
   entityManagerFactory.close();
}
```
（4）对 User 表的查询操作，代码如下。
```
@Test
public void find(){
   EntityManagerFactory entityManagerFactory = Persistence.
createEntityManagerFactory("jpa");
   EntityManager entityManager = entityManagerFactory.createEntityManager();

   User user = entityManager.find(User.class, 2);

   entityManager.close();
   entityManagerFactory.close();

   System.out.println(user.getAddress());
}
```
这里对 EntityManager 中的部分 API 进行了演示，由于本书并不专门讲解 JPA，所以对于关于 JPA 的更为高级的应用，如 JPQL、分页处理等，这里不做介绍，有兴趣的读者可以查阅相关资料或在蜗牛学院官网查看相关视频进行学习。

4.2.2 非关系型数据库的持久化实现

4.2.1 小节通过 JPA 实现了对关系型数据的添加、删除、更新和查询操作，本小节将演示如何在 Java 应用中操作 Redis 这种非关系型数据库。

Redis 是一种键/值的非关系型数据库，在项目中常用于数据的缓存，以提升数据库查询的性能，这里将通过 Java 封装的客户端 Jedis 来实现对 Redis 的操作。关于 Redis 的安装配置可以参照第 7 章的相关内容，这里不进行介绍。

具体步骤如下。

（1）开启 Redis 服务器，打开命令行窗口，执行如下命令。
```
tlhdeMBP:~ tlh$ redis-server
```
（2）创建项目并添加 Jedis 的依赖，代码如下。
```
<!-- Jedis客户端 -->
<dependency>
   <groupId>redis.clients</groupId>
   <artifactId>jedis</artifactId>
   <version>2.9.0</version>
</dependency>
```
（3）连接 Redis 服务器，代码如下。
```
@Test
public void connection(){
   Jedis jedis=new Jedis("localhost");
   System.out.println(jedis.ping());
   jedis.close();
}
```
（4）添加字符串类型数据，代码如下。
```
@Test
public void addStr(){
   Jedis jedis=new Jedis("localhost");
   jedis.set("s1", "test");

   jedis.close();
}
```

通过命令行查看添加的字符串数据。

```
tlhdeMBP:~ tlh$ redis-cli
127.0.0.1:6379> get s1
"test"
```

(5) 获取字符串数据。

```
@Test
public void getStr(){
    Jedis jedis=new Jedis("localhost");
    String s1 = jedis.get("s1");

    System.out.println(s1);

    jedis.close();
}
```

4.3　Spring Data

4.3.1　Spring Data 入门

在 4.2 节中分别采用 JPA 和 Jedis 实现了对关系型数据和非关系型数据的存储，从示例代码中可以看到不同的持久化技术，其操作方式也不尽相同。虽然它们针对的对象不一样，但是整个流程都可以抽取为 3 个阶段：获取连接、操作、释放资源。是否存在只要将基础组件封装好，只为上层应用提供操作接口即可的架构呢？运用这种架构，不管使用的是关系型数据库还是非关系型数据库，都可以以一种简洁的方式进行处理。Spring Data 就为各类数据持久化技术抽取出了基础组件并提供了相应的实现，只需要为上层应用额外提供少量的配置即可使用，其提供的丰富特性如下。

(1) 其具有强大的存储库和定制的对象映射抽象。
(2) 其基于方法名称的动态查询。
(3) 其支持透明的审计（先创建，最后更改即可）。
(4) 其支持自定义存储库实现。
(5) 其通过 JavaConfig 和定制的 XML 名称空间轻松地进行 Spring 集成。
(6) 其与 Spring MVC 控制器的高级集成满足 Web 开发中的一些特殊需求。
(7) 其支持跨存储的持久化实现。

Spring Data 为开发人员提供了丰富的特性，同时支持自定义存储库以满足特定的项目需求，并且支持跨存储方式，这样，开发人员不用再考虑在项目中针对不同的存储方式编写不同的代码。以下是 Spring Data 为开发人员提供数据存储的具体实现。

(1) Spring Data Commons：为 Spring Data 的基础组件，提供核心功能实现。
(2) Spring Data Gemfire：提供对 Gemfire 的简单配置及使用。
(3) Spring Data JPA：提供对 JPA 的实现支持。
(4) Spring Data JDBC：提供对 JDBC 的实现支持。
(5) Spring Data KeyValue：提供对键/值存储的实现支持。
(6) Spring Data REST：提供对 RESTful 的支持，将 Spring Data 的存储转换为 REST API。
(7) Spring Data Redis：提供对 Redis 的简单配置及使用。
(8) Spring Data for Apache Solr：提供对 Apache 的全文检索 Solr 的支持。

这里对 Spring Data 提供的部分主要模块做了简单介绍，除了这些官方提供的实现以外，Spring Data 社区中还提供了扩展模块，感兴趣的读者可以自行阅读 Spring Data 官方文档的介绍，这里不再做说明。

下面将选用比较常用的关系型数据存储 JPA 和非关系型数据存储 Redis 来对 Spring Data 进行讲解。

4.3.2 Spring Data JPA

前面已经对 JPA 的使用做过简单介绍，这里不再赘述。本小节将主要针对 Spring Data JPA 来讲解如何更为方便地实现数据的持久化，并利用 Spring Data JPA 对 Web 的支持满足 Web 开发的一些特定需求。

V4-1　Spring Data 简介

1. Spring Data 初探

（1）创建项目 j2e-spring-data-jpa，并添加如下依赖。

```xml
<dependencies>
    <dependency>
        <groupId>org.springframework.data</groupId>
        <artifactId>spring-data-jpa</artifactId>
        <version>2.0.0.RELEASE</version>
    </dependency>
    <dependency>
        <groupId>log4j</groupId>
        <artifactId>log4j</artifactId>
        <version>1.2.17</version>
    </dependency>
    <dependency>
        <groupId>org.slf4j</groupId>
        <artifactId>slf4j-log4j12</artifactId>
        <version>1.7.21</version>
    </dependency>
    <dependency>
        <groupId>org.hibernate</groupId>
        <artifactId>hibernate-entitymanager</artifactId>
        <version>5.2.12.Final</version>
    </dependency>
    <dependency>
        <groupId>mysql</groupId>
        <artifactId>mysql-connector-java</artifactId>
        <version>5.1.44</version>
    </dependency>
    <dependency>
        <groupId>junit</groupId>
        <artifactId>junit</artifactId>
        <version>4.12</version>
        <scope>test</scope>
    </dependency>
    <dependency>
        <groupId>org.springframework</groupId>
        <artifactId>spring-test</artifactId>
        <version>4.3.12.RELEASE</version>
        <scope>test</scope>
    </dependency>
</dependencies>
```

（2）编写配置文件，这里采用基于 Spring 提供的 JavaConfig 的方式进行配置。在项目中创建 Application 类，其内容如下。

```
@Configuration
@EnableJpaRepositories
```

```java
@EnableTransactionManagement
public class Application {

    @Bean
    public DataSource dataSource() {
        DriverManagerDataSource dataSource=new DriverManagerDataSource();
        dataSource.setUrl("jdbc:mysql://localhost:3306/jpa");
        dataSource.setDriverClassName("com.mysql.jdbc.Driver");
        dataSource.setUsername("用户名");
        dataSource.setPassword("密码");
        return dataSource;
    }

    @Bean
    public LocalContainerEntityManagerFactoryBean entityManagerFactory() {
        HibernateJpaVendorAdapter vendorAdapter = new HibernateJpaVendorAdapter();
        vendorAdapter.setGenerateDdl(true);
        vendorAdapter.setShowSql(true);

        LocalContainerEntityManagerFactoryBean factory = new LocalContainerEntityManagerFactoryBean();
        factory.setJpaVendorAdapter(vendorAdapter);
        factory.setPackagesToScan("com.woniuxy.j2e.springdata.entity");
        factory.setDataSource(dataSource());
        return factory;
    }

    @Bean
    public PlatformTransactionManager transactionManager() {
        JpaTransactionManager txManager = new JpaTransactionManager();
        txManager.setEntityManagerFactory(entityManagerFactory().getObject());
        return txManager;
    }
}
```

（3）编写 JPA 的实体类，代码如下。

```java
@Entity
@Table(name="t_users")
public class User {

    @Id
    @GeneratedValue(strategy=GenerationType.IDENTITY)
    private Integer id;
    private String userName;
    private String realName;
    private String password;
    private String address;
    private Date birthday;
    private Integer age;

    //setter\getter
}
```

（4）定义存储接口。在 4.3.1 节中了解到 Spring Data 封装了存储的底层实现，只暴露给应用层一些核心的接口，其中，Repository 是 Spring Data 的基础组件，该接口是一个泛型接口，第一个参数表示实体类的数据类型，第二个参数表示实体类中主键的数据类型。Repository 的结构如图 4-1 所示。

```
Repository<T, ID> - org.springframework.data.repository
  CrudRepository<T, ID> - org.springframework.data.repository
    PagingAndSortingRepository<T, ID> - org.springframework.data.repository
      JpaRepository<T, ID> - org.springframework.data.jpa.repository
        SimpleJpaRepository<T, ID> - org.springframework.data.jpa.repository.support
          QuerydslJpaRepository<T, ID extends Serializable> - org.springframework.data.jpa.reposit
      UserRepository - com.woniuxy.j2ee.springdata.repository
  ReactiveCrudRepository<T, ID> - org.springframework.data.repository.reactive
    ReactiveSortingRepository<T, ID> - org.springframework.data.repository.reactive
  RevisionRepository<T, ID, N extends Number & Comparable<N>> - org.springframework.data.repositor
  RxJava2CrudRepository<T, ID> - org.springframework.data.repository.reactive
    RxJava2SortingRepository<T, ID> - org.springframework.data.repository.reactive
```

图 4-1　Repository 的结构

CrudRepository：该接口为 Repository 的子接口，包含了常用的增、删、改、查操作，其源码如图 4-2 所示。

```
public interface CrudRepository<T, ID extends Serializable>
  extends Repository<T, ID> {

  <S extends T> S save(S entity);            ①

  Optional<T> findById(ID primaryKey);       ②

  Iterable<T> findAll();                     ③

  long count();                              ④

  void delete(T entity);                     ⑤

  boolean existsById(ID primaryKey);         ⑥

  // ... more functionality omitted.
}
```

图 4-2　CrudRepository 的源码

各部分主要作用如下。
① 保存实体对象。
② 通过主键查询实体对象。
③ 查询所有实体对象。
④ 查询实体对象的总数。
⑤ 删除指定的实体对象。
⑥ 通过主键判定实体对象是否存在。

V4-2　Spring Data JPA-1

PagingAndSortingRepository：该接口是 CrudRepository 接口的扩展，添加了满足 Web 开发的分页和排序功能，其源码如图 4-3 所示。

```
public interface PagingAndSortingRepository<T, ID extends Serializable>
  extends CrudRepository<T, ID> {

  Iterable<T> findAll(Sort sort);

  Page<T> findAll(Pageable pageable);
}
```

图 4-3　PagingAndSortingRepository 的源码

从图 4-1 中可以看出 JpaRepository 可扩展至 PagingAndSortingRepository，因此该接口同时兼有增、删、改、查的基础功能，以及 Web 开发的分页和排序功能，故在实际开发中，可以直接使用 JpaRepository 接口。这里先讲解 CrudRepository 接口的使用，在后面的章节中将结合 Web 环境讲解 PagingAndSortingRepository 接口的使用。

编写存储接口 UserRepository，继承自 CrudRepository 接口。

```java
public interface UserRepository extends CrudRepository<User, Integer>{

}
```

（5）进行测试，代码如下。

```java
@RunWith(SpringJUnit4ClassRunner.class)
@ContextConfiguration(classes=Application.class)
public class UserRepositoryTest {

    @Autowired
    private UserRepository userRepository;

    @Test
    public void save(){
        User user=new User();
        user.setAddress("成都");
        user.setAge(10);
        user.setRealName("离歌笑");
        user.setUserName("tlhhup");
        user.setBirthday(new Date());

        this.userRepository.save(user);
    }
}
```

由以上测试可发现，只需要编写相应的存储接口而不用编写其实现类，这就是 Spring Data 提供的功能，即通过动态代理的方式在 Spring 环境中生成该接口的实现类对象并注入 IoC 容器。这一功能只是 Spring Data 众多功能的一部分，更为丰富的特性在其方法查询上，只需要在接口中定义相应的方法，之后 Spring Data 通过解析方法名即可将其转换为相应的 Java 持久化查询语言（Java Persistence Query Language，JPQL）语句。

2．方法查询

虽然使用 Spring Data 提供的基础组件已经可以实现对数据的添加、删除、修改和查询操作，但有些情况下需要通过查询条件来过滤查询的结果，例如，要通过用户名查询用户的信息，则在数据库中需要编写相应的 where 子句，在 JPA 中需要通过 SQL 的变体 JPQL 来实现。但如果使用 Spring Data，这一需求的实现将变得尤为简单，代码如下。

```java
public interface UserRepository extends CrudRepository<User, Integer>{

    User findUserByUserName(String userName);

}
```

通过测试传入"tlhhup"参数即可查询该用户的信息，图 4-4 所示为 Spring 官方提供的接口示例。

只在接口中添加方法的声明即可完成数据的持久化，Spring Data 是如何实现的呢？其实现得益于其方法查询，只需要开发人员在定义方法时满足规定，Spring Data 即可在代理中完成相应的查询需求。Spring Data 提供了两种方式来将方法名转换为指定的查询，一种是直接通过方法名转换，另一种是由开发人员自己定义查询语句，Spring Data 提供了查询策略来控制其具体的使用方式。

```
interface PersonRepository extends Repository<User, Long> {

  List<Person> findByEmailAddressAndLastname(EmailAddress emailAddress, String lastname);

  // Enables the distinct flag for the query
  List<Person> findDistinctPeopleByLastnameOrFirstname(String lastname, String firstname);
  List<Person> findPeopleDistinctByLastnameOrFirstname(String lastname, String firstname);

  // Enabling ignoring case for an individual property
  List<Person> findByLastnameIgnoreCase(String lastname);
  // Enabling ignoring case for all suitable properties
  List<Person> findByLastnameAndFirstnameAllIgnoreCase(String lastname, String firstname);

  // Enabling static ORDER BY for a query
  List<Person> findByLastnameOrderByFirstnameAsc(String lastname);
  List<Person> findByLastnameOrderByFirstnameDesc(String lastname);
}
```

图 4-4 Spring 官方提供的接口示例

（1）查询策略。Spring Data 提供了 3 种查询策略来满足不同的需求，开发人员可以通过在 XML 文件中使用 query-lookup-strategy 属性和 JavaConfig 方式的配置类注解 Enable${store}Repositories 中的 queryLookupStrategy 来修改具体使用的查询策略。以下是 3 种具体的查询策略。

① CREATE：通过查询方法的方法名构造指定的查询。

② USE_DECLARED_QUERY：需要开发人员自己编写查询语句。

③ CREATE_IF_NOT_FOUND：这是默认的查询策略，结合了前面两种查询策略，先通过查找声明进行查询，如果没有找到，再通过方法名解析为指定的查询。

（2）方法名创建。这是最为常用的方法查询策略，Spring Data 在定义方法名时采用 find…By、read…By、query…By、count…By 和 get…By 等前缀来声明查询的方法，并且可以携带 Distinct 关键字。在进行方法名解析时，By 后面的子句将被解析为查询条件，也可以通过 And 和 Or 构建逻辑运算，这里给出查询方法定义的示例。

```
public interface UserRepository extends CrudRepository<User, Integer>{

    User findUserByUserName(String userName);

    User getUserByUserNameAndPassword(String userName,String password);

    List<User> findDistinctUserByUserName(String userName);

}
```

在前面的介绍中提到过，Spring Data 是通过在动态代理中对方法名进行解析，最终生成相应的 JPQL 语句来实现结果查询的，故在 By 后面编写的是该实体类中的属性名。Spring Data 在生成 JPQL 语句时会通过迭代的方式对 By 后面的命名进行拆分，用于生成正确的 JPQL 语句，但以下方法声明将导致这种迭代的效率低下。

```
List<Person> findByAddressZipCode(ZipCode zipCode);
```

Spring Data 是按照驼峰命名的方式对方法名进行解析的，但是以上的方法声明原本想通过 address 中的 zipCode 进行查询，通过驼峰命名的规则将被解析为 address.zip.code，此时将出现属性找不到的异常，虽然经过多次解析将得到正确的结果，但降低了效率。针对这种情况，Spring Data 推荐使用连接符（"_"）来明确查询条件，可将前面的查询方法修改如下。

```
List<Person> findByAddress_ZipCode(ZipCode zipCode);
```

上面对 Spring Data 通过方法名创建查询进行了简单介绍，有些 IDE（Integrated Development Environment，集成开发环境）中提供了更为强大的功能，可以方便地进行方法声明。除了以上讲解的规则以外，更为丰富的命名规范可以查阅 Spring Data JPA 的官网，但只要能将方法名转换为 JPQL 的方法声明就是正确的。

（3）指定查询。前面讲解了通过方法名创建的方法定义查询，但在有些情况下用户更希望通过自己编写的 JPQL 语句来实现数据的查询，或者编写原生 SQL 语句来实现查询，此时可以通过 Spring Data JPA 提供的指定查询的方法进行处理。以下代码通过指定查询的方式对前面通过用户名查询用户信息的示例进行了修改。

```java
public interface UserRepository extends CrudRepository<User, Integer>{

    User findUserByUserName(String userName);

    @Query("select u from User u where u.userName=?1")
    User findUserInfoByUserName(String userName);
}
```

这两种方法唯一的区别就是后者在方法声明时通过 @Query 注解显式声明了执行的 JPQL 语句。除了可用于声明执行的 JPQL 语句以外，@Query 注解还提供了以下属性。

① value：用于指定执行的 JPQL 或 SQL 语句。

② countQuery：用于指定执行的统计查询语句。

③ nativeQuery：布尔类型的数据，用于指定是否为原生的 SQL 查询。以下方式可以执行原生的 SQL 查询。

```java
public interface UserRepository extends CrudRepository<User, Integer>{

    User findUserByUserName(String userName);

    @Query("select u from User u where u.userName=?1")
    User findUserInfoByUserName(String userName);

    @Query(value="select * from t_users where userName=?1",nativeQuery=true)
    User findUserInfoNativeQuery(String userName);
}
```

3．修改查询

前面介绍了 Spring Data JPA 提供的方法查询，介绍了如何定义需要执行的查询，但在有些场景下同样希望执行写操作。在前面讲解的 Spring Data JPA 提供的基础组件中，CrudRepository 可以正常完成数据添加、删除和修改操作，但其提供的更新操作会出现一个问题，例如，执行以下测试。

V4-3 Spring Data JPA-2

```java
@Test
public void update(){
    User user=new User();
    user.setId(1);
    user.setAddress("北京");

    this.userRepository.save(user);
}
```

看到这个测试，细心的读者可能会有疑问，执行更新操作为何会调用 save 方法呢？通过查看执行日志，可以发现这里执行的是 update 语句，虽然调用了 save 方法，但是其会根据指定的 ID 是否在数据库中来确定是执行 update 语句还是执行 insert 语句。

通过图 4-5 所示的日志信息可以看出执行的是 update 语句，但是进行了全字段更新操作，即除了 address 属性以外，其他没有设置值的属性一并进行了更新操作。而在有些场景中是不想做全字段更新的，如修改密码时，虽然用户只修改了密码字段，但因为其他的属性没有被赋值，因此均被更新为 null 值。如何解决这个问题呢？此时需要改变执行的语句。在 Spring Data JPA 中，@Modifying 注解用于修改查询，执行写操作，如下面的 update 操作。

```
16:12:57,688 DEBUG EntityPrinter:109 - com.woniuxy.j2e.springdata.entity.User{birthday=null, realName=null, password=null, ad
16:12:57,700 DEBUG SQL:92 - update t_users set address=?, age=?, birthday=?, password=?, realName=?, userName=? where id=?
Hibernate: update t_users set address=?, age=?, birthday=?, password=?, realName=?, userName=? where id=?
```

图 4-5　日志信息

```
@Modifying
@Query("update User set password=?1 where id=2?")
int modifyingPassword(String password,int id);
```

除了在方法中添加@Modifying 注解以修改查询执行 update 以外，还要添加@Transactional 注解，因为 update 执行的是写操作，所以其需要运行在事务中。修改之后的代码如下。

```
public interface UserRepository extends CrudRepository<User, Integer>{

    User findUserByUserName(String userName);

    @Query("select u from User u where u.userName=?1")
    User findUserInfoByUserName(String userName);

    @Query(value="select * from t_users where userName=?1",nativeQuery=true)
    User findUserInfoNativeQuery(String userName);

    @Modifying
    @Transactional
    @Query("update User set password=?1 where id=?2")
    int modifyingPassword(String password,int id);

    User getUserByUserNameAndPassword(String userName,String password);

    List<User> findDistinctUserByUserName(String userName);
}
```

4. Spring Boot 与 Spring Data 的整合

通过前面的学习，相信读者可以运用 Spring Data JPA 完成各种环境下的数据持久化了。但在使用 Spring Data JPA 时需要进行一些配置，如数据源、实体管理器及事务管理器的配置，这部分重复而又必需的配置是否可以进行统一处理呢？在前面的章节中已经介绍了 Spring Boot，其自动装配的能力使其可以优雅地集成其他框架，同时，其支持 Spring Data JPA，配置类中提供了 JpaRepositoriesAutoConfiguration 和 spring-boot-starter-data-jpa 以完成对 Spring Data JPA 的快速集成。

在 Spring Boot 项目中使用 Spring Data JPA 时，只需要修改 pom.xml 文件并添加 starter，并在配置文件中添加数据源的配置即可，代码如下。

（1）添加 Spring Data JPA 的 starter。

```xml
<!-- 集成Spring Data -->
<dependency>
    <groupId>org.springframework.boot</groupId>
    <artifactId>spring-boot-starter-data-jpa</artifactId>
</dependency>
```

（2）添加 JPA 的配置信息。

```
spring:
  datasource:
    driver-class-name: org.gjt.mm.mysql.Driver
    url: jdbc:mysql:///tlh
    username: 用户名
    password: 密码
```

```
    type: com.zaxxer.hikari.HikariDataSource
  jpa:
    generate-ddl: true
    show-sql: true
```
（3）开启 JPA 配置。
```
@SpringBootApplication
@EnableTransactionManagement
@EnableJpaRepositories(basePackages = "org.tlh.examstack.module.**.repository")
public class App {

    public static void main(String[] args){
        SpringApplication.run(App.class,args);
    }

}
```
以上介绍了 Spring Data JPA，也讲解了在 Spring Boot 项目中如何使用 Spring Data JPA。除了前面讲解的关于 Spring Data 的特性以外，Spring Data JPA 还提供了更多高级的应用，如存储过程的调用、适配 Java 8 的流式结果、异步查询等，这里不再赘述，有兴趣的读者可以查询官网的相关信息。

4.3.3 Spring Data Redis

Spring Data 提供的 JPA 用于对关系型数据库进行实现，此外，Spring Data 还提供了对市面上常用的非关系型数据库的支持，本小节主要通过 Spring Data Redis 来讲解 Spring Data 对 Redis 等键/值存储的非关系型数据库的支持。

1. 创建项目 j2e-spring-data-redis 并添加依赖

具体代码如下。
```xml
<dependencies>
    <dependency>
        <groupId>log4j</groupId>
        <artifactId>log4j</artifactId>
        <version>1.2.17</version>
    </dependency>
    <dependency>
        <groupId>org.slf4j</groupId>
        <artifactId>slf4j-log4j12</artifactId>
        <version>1.7.21</version>
    </dependency>
    <dependency>
        <groupId>redis.clients</groupId>
        <artifactId>jedis</artifactId>
        <version>2.9.0</version>
    </dependency>
    <dependency>
        <groupId>org.springframework.data</groupId>
        <artifactId>spring-data-redis</artifactId>
        <version>2.0.3.RELEASE</version>
    </dependency>
    <dependency>
        <groupId>junit</groupId>
        <artifactId>junit</artifactId>
        <version>4.12</version>
        <scope>test</scope>
    </dependency>
```

```xml
<dependency>
    <groupId>org.springframework</groupId>
    <artifactId>spring-test</artifactId>
    <version>5.0.3.RELEASE</version>
    <scope>test</scope>
</dependency>
</dependencies>
```

2. 配置会话

在项目中创建 Application 类，配置 RedisConnectionFactory 对象，这里采用 Jedis 作为客户端并连接 Redis。配置 Spring Data Redis 提供的模板 StringRedisTemplate 以实现对 Redis 的操作，其内容如下。

```java
@Configuration
public class Application {

    @Bean
    public RedisConnectionFactory redisConnectionFactory() {
        JedisConnectionFactory connectionFactory = new JedisConnectionFactory();
        return connectionFactory;
    }

    @Bean
    public StringRedisTemplate stringRedisTemplate(){
        StringRedisTemplate redisTemplate=new StringRedisTemplate();
        redisTemplate.setConnectionFactory(redisConnectionFactory());

        return redisTemplate;
    }
}
```

Spring Data Redis 中内置的模板类封装了对 Redis 底层的操作，有以下两种模板类。

（1）RedisTemplate：封装了对 Redis 底层操作的实现，提供了简单的 API 以存储数据。

（2）StringRedisTemplate：针对字符串类型的数据操作的模板对象，其为 RedisTemplate 的子类，其中指定了采用 StringRedisSerializer 作为存储的序列化组件。

3. 测试

确保 Redis 服务已经开启，编写测试类添加字符串类型的数据。

（1）添加字符串类型的数据，代码如下。

```java
@RunWith(SpringJUnit4ClassRunner.class)
@ContextConfiguration(classes=Application.class)
public class RedisTemplateTest {

    @Autowired
    private StringRedisTemplate stringRedisTemplate;

    @Test
    public void setStr(){
        ValueOperations<String, String> opsForValue = this.stringRedisTemplate.opsForValue();
        opsForValue.set("s1", "get");
    }
}
```

运行测试代码，成功之后登录 Redis 客户端，可以查看刚添加的数据，如图 4-6 所示。

```
127.0.0.1:6379> get s1
"get"
127.0.0.1:6379>
```

图 4-6　查看刚添加的数据

（2）在已有的 key 中追加字符串数据，代码如下。

```
@Test
public void append(){
    ValueOperations<String, String> opsForValue = this.stringRedisTemplate.opsForValue();
    Integer append = opsForValue.append("s1", " a message");
    System.err.println(append);
}
```

（3）查看指定 key 中存储的数据，代码如下。

```
@Test
public void getStr(){
    ValueOperations<String, String> opsForValue = stringRedisTemplate.opsForValue();
    String string = opsForValue.get("s1");
    System.out.println(string);
}
```

4. 在 Spring Boot 中集成 Redis

在 Spring Boot 项目中集成 Redis 的步骤和集成 JPA 的步骤类似，只需要在项目中添加相应的 starter 依赖即可，这里不再赘述。

本章先采用传统方式分别实现了对关系型数据库和非关系型数据库的持久化，又通过引入 Spring Data 来简化持久化操作。通过对本章的学习，在以后的项目中，读者不用为采用不同的持久化技术而配置各种连接对象，而可以把更多的精力放在业务实现上。对其他 Spring Data 实现有兴趣的读者，可以自行查阅官方文档。

V4-4　Spring Data Redis

第5章

模板引擎

学习目标

（1）了解常用的模板引擎。
（2）掌握FreeMarker的使用及其与Spring Boot的整合。

本章导读

■在做 Web 开发的过程中会涉及 PC 展示，即将需要展现的数据显示到浏览器上，这就是通常所说的 View 层。在数据展现的过程中，一般会用到一些模板引擎来简化数据绑定过程，如第 1 章学习的 JSP 就属于一种模板引擎，其在使用过程中会借助 JSTL 和 EL 表达式来简化数据绑定过程。本章将讲解另一种常用的模板引擎——FreeMarker 在开发中的应用，并结合 Spring Boot 来做简单的演示。

5.1 常用模板引擎

5.1.1 模板引擎

模板引擎（这里特指用于 Web 开发的模板引擎）是为了使用户界面与业务数据（内容）分离而产生的，它可以生成特定格式的文档，例如，用于网站的模板引擎就会生成一个标准的 HTML 文档。前端人员定义好模板，后端开发人员给定相应的数据，再结合模板引擎进行数据的绑定及视图渲染后，即可将成品展现给用户。模板引擎的结构如图 5-1 所示。

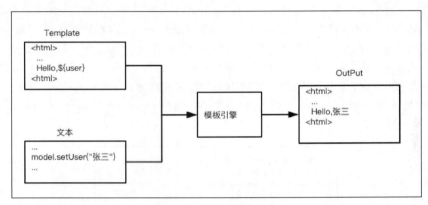

图 5-1　模板引擎的结构

通过引入模板引擎可以解决前后分离的问题，这里整理了在实际开发过程中常用的模板引擎，分别介绍如下。

1. JSP

JSP 技术使用 Java 编程语言编写类 XML 的 tags 和 scriptlets 来封装产生动态网页的处理逻辑。网页能通过 tags 和 scriptlets 访问存在于服务端的资源的应用逻辑。JSP 将网页逻辑与网页设计的显示分离，支持可重用的基于组件的设计，使基于 Web 的应用程序的开发变得迅速和容易。JSP 是一种动态页面技术，它的主要目的是将表示逻辑从 Servlet 中分离出来。

2. FreeMarker

FreeMarker 是一个基于模板生成文本输出的通用工具，使用纯 Java 编写。FreeMarker 是为 Java 程序开发人员提供的一个开发包，其使用 MVC 模式的动态页面的设计构思使得程序开发人员可以将前端设计师从程序设计中分离出来。

3. Thymeleaf

Thymeleaf 是网站或独立应用程序的新式服务端 Java 模板引擎，可以执行 HTML、XML、JavaScript、CSS 甚至纯文本模板。

4. Velocity

Velocity 是一个基于 Java 的模板引擎。它允许任何人使用简单而强大的模板语言来引用 Java 代码中定义的对象。

5. Beetl

Beetl 是一个国产的模板引擎，它是新一代的模板引擎，其功能强大，性能良好，易学易用。

以上是对日常开发中常用到的模板引擎的介绍，在具体项目中使用哪种模板引擎

V5-1　模板引擎简介

还要视情况而定，同时，关于各个模板引入性能的对比也是在选型时要考虑的，这部分内容读者可以自行参考网络中的资料。

5.1.2　Spring Boot 对模板引擎的支持

在第 3 章中使用 Spring Boot 快速搭建了 Spring MVC 的开发环境，但只是简单地以 JSON 数据格式展现给了用户，没有通过模板引擎或 HTML 的方式来展现用户数据，那么 Spring Boot 中支持的模板引擎有哪些呢？

从 Spring 官方文档介绍中可知，其除了支持 REST 的 Web Services 以外，还可以通过 Spring MVC 开发动态 HTML，所以只要是 Spring MVC 支持的模板引擎，理论上 Spring Boot 都支持。Spring MVC 对 5.1.1 小节中介绍的模板引擎都是支持的，但在 Spring Boot 项目中，不推荐使用 JSP 作为模板引擎，主要原因在于 Spring Boot 支持内嵌 Servlet 容器，而内嵌的 Servlet 容器对 JSP 的支持是不太友好的。

Spring Boot 中包含的自动装配的模板引擎包括：FreeMarker、Groovy、Thymeleaf、Mustache。在以往的 Web 开发中，与界面相关的文件都是存储在 webapp 或 webcontent 目录中的，而在基于 Spring Boot 的项目中会有所区别，所有和界面相关的模板数据都需要存储在 src/main/resources/templates 目录中，其他的静态资源（JavaScript、图像、CSS）需要存储在 src/main/resources/static 目录中，具体的文件结构可参考图 5-2。

图 5-2　具体的文件结构

5.2　FreeMarker 引擎

5.2.1　了解 FreeMarker

5.1 节对常用的模板引擎做了简单介绍，本节主要围绕 FreeMarker 来学习一种新的模板引擎。

FreeMarker 是免费的，其基于 Apache 许可证 2.0 版本发布，是一种基于模板及要改变的数据，并生成输出文本（HTML 网页、电子邮件、配置文件、源代码等）的通用工具。FreeMarker 不是面向最终用户的，而是一个 Java 类库，是一款开发人员可以嵌入在其所研发产品中的组件。

FreeMarker Template Language 是简单的、专用的语言，不是 PHP 那样成熟的编程语言。这就意味着要准备数据并通过编程语言来显示，例如，在数据库查询和业务运算之后，交由模板将准备好的数据进行展示，在模板中可以专注于如何展示数据，而在模板之外可以专注于要展示什么数据，使得前后分离，以提高开发效率。FreeMarker 的结构如图 5-3 所示。

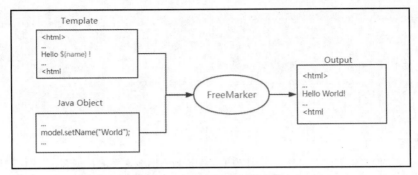

图 5-3　FreeMarker 的结构

这通常被称为 MVC 模式，对于动态网页开发来说，这是一种特别流行的模式。它从开发人员中分离出网页设计师。网页设计师无需面对模板中的复杂逻辑，在没有程序开发人员来修改或重新编译代码时，也可以修改页面的样式。

FreeMarker 最初是被用来在 MVC 模式的 Web 开发框架中生成 HTML 页面的，它没有被绑定到 Servlet、HTML 或任意与 Web 相关的事物上，所以它可以用于非 Web 应用环境中。

下面通过 Spring Boot 结合 FreeMarker 搭建一个入门案例。

1．创建项目并添加依赖

具体代码如下。

```xml
<parent>
    <groupId>org.springframework.boot</groupId>
    <artifactId>spring-boot-starter-parent</artifactId>
    <version>1.4.7.RELEASE</version>
</parent>
<dependencies>
    <dependency>
        <groupId>org.springframework.boot</groupId>
        <artifactId>spring-boot-starter-web</artifactId>
    </dependency>
    <dependency>
        <groupId>org.springframework.boot</groupId>
        <artifactId>spring-boot-starter-freemarker</artifactId>
    </dependency>
</dependencies>
```

2．创建 Web 目录

在项目的 src/main/resources 目录中分别创建 static 和 templates 文件夹，Web 目录的结构如图 5-4 所示。

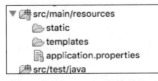

图 5-4　Web 目录的结构

3．编写启动类

在项目的 src/main/java 目录中创建 com.woniuxy.ssf 包，并在此包中创建 Application 类，其内容和第 3 章学习的 Spring Boot 的启动类内容一致，代码如下。

```java
@SpringBootApplication
public class Application {

    public static void main(String[] args) {
        SpringApplication.run(Application.class, args);
    }

}
```

4. 编写模板

在 templates 文件夹中创建 hello.ftl（.ftl 为 freeMarker 模板文件的扩展名）文件，其内容如下。

```html
<!DOCTYPE html>
<html>
<head>
<meta charset="UTF-8">
<title>FreeMarker入门</title>
</head>
<body>
  Hello,${user}
</body>
</html>
```

5. 编写处理器

在项目的 src/main/java 目录中创建 com.woniuxy.ssf 包，并在此包中创建子包 controller，创建 HelloController 类，用于在 model 中设置数据，内容如下。

```java
@Controller
public class HelloController {

    @RequestMapping("/")
    public String hello(Model model){
        model.addAttribute("user", "张三");
        return "hello";
    }

}
```

6. 测试

启动配置类，打开浏览器并访问 http://localhost:8080，即访问 hello.ftl 页面，其中展示了在处理器中设置的数据，如图 5-5 所示。

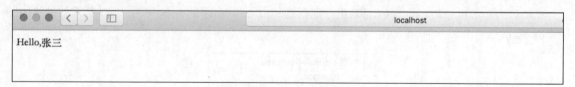

图 5-5 访问 hello.ftl 页面

回顾整个过程，数据展示和数据已经完全分离了，后端开发人员不用关注前端数据如何展示，只需要设置相应的数据即可，剩下的都交由模板引擎来实现，所以 FreeMarker 中有这样一种说法："模板+数据模型=输出"。

关于模板的知识将在之后的章节中进行更为详细的介绍，这里先对模型数据的格式做简单讲解。从上面的示例程序中可以看到，这里只是向 model 中添加了一个属性名为"user"的属性，这样就可以在前端页面进行展示，而 model 的结构本质上类似于 Java 中的 Map，属于一种键/值对的树形的数据结构。model 的数据结构如图 5-6 所示。

```
DATA MODEL
(root)
  |
  +- animals
  |   |
  |   +- mouse
  |   |   |
  |   |   +- size = "small"
  |   |   |
  |   |   +- price = 50
  |   |
  |   +- elephant
  |   |   |
  |   |   +- size = "large"
  |   |   |
  |   |   +- price = 5000
  |   |
  |   +- python
  |       |
  |       +- size = "medium"
  |       |
  |       +- price = 4999
  |
  +- message = "It is a test"
  |
  +- misc
      |
      +- foo = "Something"
```

图 5-6　model 的数据结构

5.2.2　FreeMarker 类型

V5-2　在 Spring Boot 中使用模板引擎

5.2.1 小节对模板的数据格式做了简单介绍，但是数据有其相应的数据类型，那么 FreeMarker 中的数据都存在哪些类型呢？注意，这里说的类型和之前说的数据类型是有所区别的，可简单理解为类型包含数据类型。以下为 FreeMarker 所支持的类型，其中比较特殊的类型为子程序。

1. 标量

标量是最基本也最简单的类型，它所包括的数据类型如下。

（1）字符串：表示简单的文本数据，如上方入门案例中的"张三"。如果想在模板中直接给出字符串值，而不是数据模型中的变量，那么将文本内容写在引号内（单引号和双引号都可以）即可，如'张三'或"张三"。

（2）数值：和 Java 中的数值类型的数据所表示的意思相同，包括整型和浮点型的数据。

（3）布尔值：布尔值代表了逻辑上的对或错（是或否）。在模板中，可以使用保留字 true 和 false 来指定布尔值。

（4）日期：日期变量可以存储和日期/时间相关的数据，共有以下 3 种情况。

① 日期：精确到天的日期，没有时间部分，如 April 4, 2003。

② 时间：精确到毫秒，没有日期部分，如 10:19:18 PM。

③ 日期-时间（也被称为"时间戳"）：有时间和日期两部分，如 April 4, 2003 10:19:18 PM。

2. 容器

该类型的数据包括 Java 中的数组和集合类型的数据，用于存储多个值。容器类型包含的数据类型如下。

（1）哈希表：该类型的数据类似于 Java 中 Map 类型的数据，属于键/值对格式。每个子变量都可以通过一个唯一的名称来查找，这个名称是不受限制的字符串。哈希表并不确定其中子变量的顺序，也就是说，没有第一个子变量、第二个子变量这样的说法，变量仅仅是通过名称来访问的。

（2）序列：该类型的数据类似于 Java 中的数组，每个子变量都通过一个整数来标识。第一个子变量的标识符是 0，第二个子变量的标识符是 1，第三个子变量的标识符是 2，以此类推，且子变量是有顺序的。这些变量的顺序通常被称为 index（索引）。但是它和 Java 中的数组类型也存在一些差异，如 FreeMarker 的序列中的子变量类型不需要完全一致。

（3）集合：集合可以理解为一个有限制的序列。用户无法获取集合的大小，也无法通过索引取出集合中的子变量，但是可以通过 list 指令来遍历集合中的数据。

3. 子程序

子程序主要用于包含方法和函数以及用户自定义指令，其类似于 Java 中提供的内置函数和开发人员定义的函数。

（1）方法和函数：当一个值是方法或函数的时候，其可以计算其他值，结果取决于传递给它的参数。例如，定义以下模板，其最终的输出结果是调用 avg 函数执行之后的结果。

```
The average of 3 and 5 is: ${avg(3, 5)}
The average of 6 and 10 and 20 is: ${avg(6, 10, 20)}
```

（2）用户自定义指令：用户自定义指令是一种子程序，是一种可以复用的模板代码段。

5.2.3　FreeMarker 模板

前面学习了 FreeMarker 中模型的数据类型，下面将结合具体的示例来学习编写模板以及使用模板中常用指令的方法。

1. 模板构成

在讲解如何编写模板之前，先来了解一下模板主要由哪些部分组成，以从全局的角度看清模板的本质。模板主要由以下 4 部分构成。

（1）文本：文本会直接原样输出，展示在界面中，不做任何解析。

（2）插值：插值的输出会被相应的计算值替换。

（3）FTL 标签：其和 HTML 标签类型的唯一区别在于，FTL 标签是由 FreeMarker 提供的。

（4）注释：其和 HTML 的注释相似，区别在于 FreeMarker 中的注释是被包含在以 "<#--" 开头和以 "-->" 结尾的标签对中的，这部分内容不会在界面中被展示出来，如下代码所示。

```
<body>
    <#-- 展示用户信息 -->
    <table>
        <thead>
            <tr>
                <th>姓名</th>
                <th>体重</th>
                <th>年龄</th>
                <th>生日</th>
                <th>是否可用</th>
            </tr>
        </thead>
        <tbody>
```

```
        <#list users as user>
        <tr>
            <td>${user.name}!</td>
            <td>${user.weight}</td>
            <td>${user.age}</td>
            <td>${user.birthday}</td>
            <td>${user.actived?then('可用','禁用')}</td>
        </tr>
        </#list>
    </tbody>
  </table>
</body>
```

2. 用户信息展示

下面通过用户信息展示的例子来讲解如何编写模板，包括指令、表达式和插值的使用。

（1）编写用户实体类，代码如下。

```java
public class User {

    private Integer id;
    private String name;
    private Double weight;
    private Integer age;
    private Boolean actived;
    private Date birthday;
}
```

（2）编写 UserService 类以模拟列表数据，代码如下。

```java
@Service("userService")
public class UserService {

    private static List<User> users;

    public List<User> findAll(){
        if(users==null){
            users=new ArrayList<User>();
            User user=null;
            for(int i=0;i<5;i++){
                user=new User();
                user.setId(i+1);
                user.setActived(i%2==0);
                user.setBirthday(new Date());
                user.setName("这个是第"+(i+1)+"个数据");
                user.setWeight(Math.random()*200);
                user.setAge(25);

                users.add(user);

                user=null;
            }
        }
        return users;
    }
```

}

（3）编写模板，代码如下。

```html
<html>
<head>
<meta charset="UTF-8">
<title>用户列表</title>
</head>
<body>
    <#-- 展示用户信息 -->
    <table>
        <thead>
            <tr>
                <th>姓名</th>
                <th>体重</th>
                <th>年龄</th>
                <th>生日</th>
                <th>是否可用</th>
            </tr>
        </thead>
        <tbody>
            <#list users as user>
            <tr>
                <td>${user.name}!</td>
                <td>${user.weight}</td>
                <td>${user.age}</td>
                <td>${user.birthday?date }</td>
                <td>${user.actived?then('可用','禁用')}</td>
            </tr>
            </#list>
        </tbody>
    </table>
</body>
</html>
```

（4）编写处理器，代码如下。

```java
@Controller
public class UserController {

    @Autowired
    private UserService userService;

    @RequestMapping("/list")
    public String list(Model model){
        model.addAttribute("users", this.userService.findAll());
        return "user";
    }

}
```

（5）进行测试，通过浏览器可以看到展示出来的用户信息，如图 5-7 所示。

图 5-7 用户信息

（6）说明。这个示例中包含了 FreeMarker 的大部分内容，这里对模板中使用到的知识点做如下说明。

<#list users as user>：这是 FreeMarker 中的指令，类似于 JSTL 中的 foreach 标签，用于对列表类型的数据进行遍历，其具体语法如下。

```
<#list sequence as item>
    Part repeated for each item
<#else>
    Part executed when there are 0 items
</#list>
```

${user.name}!：这是一个插值表达式，用于将转换后的值展示在界面中。注意后面的"!"，这里使用了 FreeMarker 中对不存在的值的处理方式。因为 name 为字符串类型的数据，所以如果没有设置值，则模板引擎在进行渲染的时候会报错，这时可以通过"!"来设置默认值，具体语法如下。

```
unsafe_expr!default_expr 或 unsafe_expr! 或 (unsafe_expr)!default_expr 或 (unsafe_expr)!
```

${user.birthday?date}：这里调用了 FreeMarker 中的函数来对日期类型的数据进行格式化，以上只显示日期。FreeMarker 中提供了 date、time 和 datetime 3 个内置函数来对日期类型的数据进行格式化。

${user.actived?then('可用','禁用')}：这里使用了布尔值的内置函数，其语法类似于 Java 程序中的三元表达式，具体语法如下。

```
booleanExp?then(whenTrue, whenFalse)
```

除了以上示例中使用到的指令、表达式及插值以外，还有一个比较常用的 if 指令，其语法如下。

```
<#if condition1>
   ...
<#elseif condition2>
   ...
<#elseif condition3>
   ...
...
<#else>
   ...
</#if>
```

通过本章的学习，相信读者已经掌握了如何在项目中运用 FreeMarker，虽然本章讲解的内容有限，但是这些都是实际项目开发中最应掌握的内容，为学习更详细的知识点，读者可以查询 FreeMarker 官网中相应的文档，但在实际开发中，编者建议遇到问题之后再查询官网的资料，这有助于快速掌握一种新技术。

V5-3　FreeMarker 应用

第6章

Shiro权限管理

学习目标

（1）了解权限操作。
（2）掌握基于Shiro的Web程序权限控制。
（3）掌握Shiro与Spring的整合应用。

本章导读

■前面为读者介绍了企业开发中的实用技术，也是目前较为流行的技术。众所周知，大部分系统需要权限的控制，第1章即介绍了使用Servlet完成权限控制的示例。虽然在企业开发时，也可以选择这样的形式完成权限的控制，但如果要完成更加复杂的权限控制，工作量是巨大的。为了更加系统、完善地进行权限控制，可以选用功能较为齐全的框架，而Shiro就是这种框架，Shiro除了可以完成权限控制之外，还可以对项目进行安全管理。

6.1 Shiro 简介

Shiro 是 Apache 的一个功能强大且易用的安全框架,其采用 Java 语言编写,可以非常轻松地执行身份认证、授权、密码加密等功能的管理。Shiro 的 API 易于理解,在开发中,可以非常方便地完成复杂的功能。Shiro 可广泛应用于各种应用程序中,而不是只局限于 Web 应用。当然,开发人员的任务是在 Web 程序中融入 Shiro 进行权限、安全管理。首先,有必要对权限管理进行全面认识。

权限管理通常是所有系统都会设置的安全规则,或者称为安全策略。在系统中,用户可以访问数据,但不是每个用户都可以访问所有数据,如当数据不属于该用户或用户没有查看权限等时则不可访问这部分数据。而实际项目的权限设计并不是直接为用户设定权限,而是先为用户赋予角色,再为角色设置可访问资源,不同的角色能访问的资源不同,可以通过为用户添加/删除角色以达到权限控制的目的。

下面举几个例子,以使读者对权限管理有更深的体会。

(1) 在很多应用中有会员机制,并且有对应的积分制度,随着积分的累积,会员的等级会提升,而会员等级的提升意味着更多功能的开放。例如,有用户"张三",为其赋予"人力资源经理"角色,其便具有了"查询员工""添加员工""修改员工""删除员工"的权限,而如果去掉"张三"的"人力资源经理"角色,那么"张三"便不能再进行这些操作。

(2) 在总公司和分公司中,"李四"是总公司"人力资源经理"角色,而"王五"是贵州分公司"人力资源经理"角色,那么"李四"可以对公司中所有的员工进行操作,而"王五"只能对贵州分公司的员工进行操作。

可以看到,权限是对用户的操作做出限制,这样做的目的是保证系统的安全性、完整性,通过 Shiro,无论是小型的管理项目还是大型的互联网项目,都可以方便地完成这些功能。

在 Shiro 设计的架构中,包含以下 3 个非常重要的组件。

(1) Subject:可以理解为当前用户,但 Shiro 对 Subject 的定义更为广泛,可以是人、进程、守护进程、计划任务等,泛指当前和软件交互的事物。

(2) SecurityManager:Shiro 的核心机制,是整个 Shiro 框架的控制器,通过协调其他组件而完成对用户的认证和授权,也称为安全管理器。

(3) Realm:Realm 是 Shiro 获取用户、权限等数据的组件,例如,若想要从数据库中获取用户数据,可以使用 JdbcRealm 组件。Shiro 框架中提供了内置的 Realm,也可以自定义 Realm 以扩展对特定功能的支持。

Shiro 的架构如图 6-1 所示。

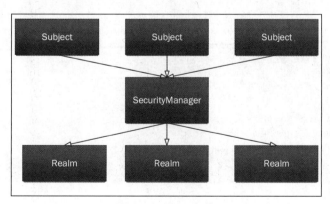

图 6-1 Shiro 架构

从图 6-1 中可以看到,在一个 Shiro 的程序架构中,Subject 对象可以有多个,表示程序中的多个用户；Realm 也可以有多个，表示不同的数据源。

当然，Shiro 框架中不仅仅包含前面提到的组件，还包含以下组件。

（1）Authenticator：用户认证管理器，用于对用户身份进行验证，通常用于处理登录逻辑，其调用 Realm 来完成对用户登录信息的判断。

（2）Authorizer：权限管理器，此对象用于完成用户的访问控制，前面讲到的各种权限方面的案例都可以通过此对象完成。

（3）Cryptography：加密技术，在安全方面，信息加密通常是必需的，而 Shiro 提供了一套易于理解、使用简单的加密 API，对常用的散列算法、对称加密算法等提供了支持。

Shiro 提供的 SecurityManager 是以模块化的形式实现的，所有的功能都通过子组件完成，所以，使用 Shiro 可以方便地完成功能的插拔。由于 SecurityManager 支持 JavaBean，因此可以使用与 JavaBean 兼容的机制为 SecurityManager 注入定制化的组件。例如，Shiro 可以方便地与 Spring 进行整合使用。

6.2 用户认证

身份认证是应用程序安全保证的前提，就像人们买票坐车一样。出于安全的考虑，需要对身份进行验证，而通常身份验证采用身份证的形式，对每个人做一个唯一的标识，即身份证号码，如果认证通过，则可以进行购票、乘车、住宿等操作。

类比到系统中，用户若想要使用系统的功能，就必须对系统出示身份证明，即证明其有操作的权限。

在 Shiro 中，对用户进行验证需要用户提供 principals 和 credentials，即身份和证明。下面通过一个简单的例子加以说明。

1. 项目准备

本项目使用 Maven 进行项目的管理，对 Maven 操作不熟悉的读者可参考第 1 章有关内容。创建 Maven 项目，在 pom.xml 文件中添加如下依赖。

```xml
<dependencies>
    <!-- JUnit单元测试 -->
    <dependency>
        <groupId>junit</groupId>
        <artifactId>junit</artifactId>
        <version>4.9</version>
    </dependency>
    <!-- 日志记录 -->
    <dependency>
        <groupId>commons-logging</groupId>
        <artifactId>commons-logging</artifactId>
        <version>1.1.3</version>
    </dependency>
    <!-- Shiro核心 -->
    <dependency>
        <groupId>org.apache.shiro</groupId>
        <artifactId>shiro-core</artifactId>
        <version>1.2.2</version>
    </dependency>
</dependencies>
```

2. 用户身份准备

本例使用了 Shiro 提供的 INI 机制，即使用 INI 文件对身份进行设置，在 resources 目录中创建文件 shiro.ini，内容如下。

```
[users]
#INI是Shiro提供的初始化数据的一种方式,以文件的形式对数据进行配置
#users标签用于设置用户信息,格式为username=password
zhang=123
wang=123
```

3. 测试

在 test/java 目录中创建类 com.woniu.test.ShiroTest,添加测试代码,本章使用 JUnit 方式进行测试。

```
@Test
public void testLogin() {
    //通过配置文件初始化SecurityManager对象
    Factory<SecurityManager> f = new IniSecurityManagerFactory("classpath:shiro.ini");
    //获取SecurityManager对象
    SecurityManager sm = f.getInstance();
    //将SecurityManager对象绑定到SecurityUtils上
    SecurityUtils.setSecurityManager(sm);
    //得到Subject
    Subject subject = SecurityUtils.getSubject();
    //创建用户名/密码验证Token
    UsernamePasswordToken token =
            new UsernamePasswordToken("zhang", "123");
    //判断用户是否已经登录
    if(subject.isAuthenticated()) {
        System.out.println(subject.getPrincipal()+" 已登录");
    } else {
        //进行登录
        subject.login(token);
        System.out.println(
                subject.getPrincipal()+" login successful");
    }
    //退出登录
    subject.logout();
    if(subject.getPrincipal()==null) {
        System.out.println("logout!!");
    }
}
```

这个测试案例流程很简单,却包含基本的 Shiro 处理过程,即收集用户信息→使用 Subject 进行登录验证→调用退出。不管多复杂的 Shiro 应用,其基本的执行流程也是在此基础之上进行扩展的。Shiro 的验证流程如图 6-2 所示。

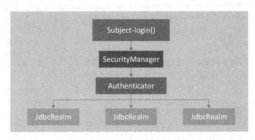

图 6-2 Shiro 的验证流程

实际上,Shiro 的 Realm 是获取数据源的组件,在本例中为了方便测试,使用了配置文件对数据进行设置,但是在实际项目开发中,数据通常存储在数据库中,可以遵照 Shiro 的规范,使用 JdbcRealm

进行身份认证。

（1）导入 JAR 包。在 pom.xml 文件中添加如下依赖。

```xml
<!-- JDBC -->
<dependency>
    <groupId>mysql</groupId>
    <artifactId>mysql-connector-java</artifactId>
    <version>5.1.25</version>
</dependency>
<!-- 数据库连接池 -->
<dependency>
    <groupId>com.alibaba</groupId>
    <artifactId>druid</artifactId>
    <version>1.0.29</version>
</dependency>
```

（2）创建数据库。通过 MySQL 创建数据库，并创建 3 张表，分别是 users（用户表）、user_role（用户角色表）和 role_permission（角色权限表），并在表中插入数据，代码如下。

```sql
CREATE DATABASE shiro;
USE shiro;
#创建用户角色表
CREATE TABLE 'user_role' (
  'id' int(11) NOT NULL AUTO_INCREMENT PRIMARY KEY COMMENT '主键',
  'rolename' varchar(20) DEFAULT NULL COMMENT '角色名称'
) ENGINE=InnoDB AUTO_INCREMENT=4 DEFAULT CHARSET=utf8;
#插入测试数据
INSERT INTO user_role(rolename) values('system'),('admin'),('other');

#创建用户表
CREATE TABLE 'users' (
  'id' int(11) NOT NULL AUTO_INCREMENT PRIMARY KEY COMMENT '用户主键',
  'username' varchar(20) NOT NULL COMMENT '用户名',
  'password' varchar(20) NOT NULL COMMENT '密码',
  'role_id' int(11) DEFAULT NULL COMMENT '外键关联role表'
) ENGINE=InnoDB AUTO_INCREMENT=4 DEFAULT CHARSET=utf8;
#插入测试数据
INSERT INTO users(username,password,role_id) values('woniu1','123456','1'),
('woniu2','123456','2'),('woniu3','123456','3');

#创建角色权限表
CREATE TABLE 'role_permission' (
  'id' int(11) NOT NULL AUTO_INCREMENT PRIMARY KEY COMMENT '主键',
  'permissionname' varchar(50) NOT NULL COMMENT '权限名',
  'role_id' int(11) DEFAULT NULL COMMENT '外键关联role'
) ENGINE=InnoDB AUTO_INCREMENT=3 DEFAULT CHARSET=utf8;
#插入测试数据
INSERT INTO role_permission(permissionname,role_id) VALUES('system:*',1),('user:*',2),
('other',3);
```

（3）创建 INI 文件。创建 shiro-jdbc.ini 文件，配置 Realm、连接数据库环境等。

```ini
jdbcRealm=org.apache.shiro.realm.jdbc.JdbcRealm
dataSource=com.alibaba.druid.pool.DruidDataSource
dataSource.driverClassName=com.mysql.jdbc.Driver
dataSource.url=jdbc:mysql://localhost:3306/shiro
dataSource.username=root
dataSource.password=root
jdbcRealm.dataSource=$dataSource
```

```
securityManager.realms=$jdbcRealm
```
（4）进行测试。
```java
@Test
public void test3() {
    IniSecurityManagerFactory factory = new 
        IniSecurityManagerFactory("classpath: shiro_jdbc.ini");
    SecurityManager sm = factory.getInstance();
    SecurityUtils.setSecurityManager(sm);
    Subject subject = SecurityUtils.getSubject();

    UsernamePasswordToken token = 
                new UsernamePasswordToken("zhang", "123");
    try {
        subject.login(token);
        System.out.println("登录成功");
    } catch (Exception e) {
        e.printStackTrace();
        System.out.println("登录失败");
    }
    subject.logout();
}
```

6.3 用户授权

用户授权也称访问控制，即设置用户可以访问什么数据/资源。在授权中，涉及以下几个对象。

1．主体

在 Shiro 中，主体即 Subject，可以认为是用户。

2．资源

资源指在系统中可以访问的数据，如 JSP 页面、某个方法等。用户不能随意访问这些资源，必须具有相应的权限才能访问。

3．角色

一个用户具有多个角色，而一个角色可具有多个权限，通过用户-角色-权限可找到用户可以进行操作的集合。

4．权限

权限表示用户在系统中是否具有操作某个资源的权力。对资源的控制不是反映到底由谁去执行这个操作，而是反映角色是否具有操作的能力，通过为用户赋予角色，可关联到用户的权限。用户与角色是一对多关系，即一个用户可能具有多个角色，而角色与权限也是一对多关系，即一个角色可能具有多个权限。

下面通过代码来展示用户角色基于资源的访问案例。

（1）配置 INI 文件。创建文件 shiro_permissions.ini，内容如下。

```ini
[users]
#配置用户-角色，配置规则为username=password,角色1,角色2
zhang=123,role1,role2
wang=123,role1
[roles]
#配置角色-资源权限，配置规则为角色=资源权限
#如角色1,为其赋予user模块的create(新建)、update(更新)权限
#如此,在对user模块进行操作时,就可以判断用户是否具有相应的权限而决定是否放行
role1=user:create,user:update
```

```
role2=user:create,user:delete
```

（2）进行测试。为了方便测试，在代码中加入JUnit断言功能，代码如下。

```java
@Test
public void testPermission() {
    //获取Subject,登录
    Factory<SecurityManager> factory = new
        IniSecurityManagerFactory("classpath:shiro_permissions.ini");
    SecurityManager sm = factory.getInstance();
    SecurityUtils.setSecurityManager(sm);
    Subject subject = SecurityUtils.getSubject();
    UsernamePasswordToken token =
            new UsernamePasswordToken("zhang", "123");
    subject.login(token);
    //判断是否具有单个操作权限
    Assert.assertTrue(subject.isPermitted("user:create"));
    //判断是否具有所有的操作权限，全部具有时返回true
    Assert.assertTrue(
            subject.isPermittedAll("user:update","user:delete"));
    //检查是否具有某权限，没有则抛出异常
    subject.checkPermission("user:create");
    subject.checkPermissions(
            "user:update","user:delete","user:view");
}
```

这里通过单个资源-单个权限的形式进行权限匹配，为了方便用户的使用，Shiro提供了更多方式来匹配权限，具体列举如下。

（1）单个资源-单个权限：role=user:create。

（2）单个资源-多个权限：role1=menu:create,get。

（3）单个资源-全部权限：role2=menu:create,get,update,delete。

（4）所有资源-单个权限：*:create。

（5）对资源的操作一般为增、删、查、改，所以全部权限可以为create、get、update、delete。如果具有全部权限，则可以使用通配符的形式：menu:*或menu。

Shiro提供的checkPermission、checkPermissions用于对用户的权限进行授权验证，实际上，Shiro也提供了对用户的角色进行授权验证的方式。

（1）创建INI文件，内容如下。

```
[users]
#测试案例中仅对用户赋予角色
zhang=123,role1,role2
wang=123,role1
```

（2）进行测试，代码如下。

```java
@Test
public void testRole() {
    Factory<SecurityManager> factory = new
        IniSecurityManagerFactory("classpath:shiro_role.ini");
    SecurityManager sm = factory.getInstance();
    SecurityUtils.setSecurityManager(sm);
    Subject subject = SecurityUtils.getSubject();
    UsernamePasswordToken token =
            new UsernamePasswordToken("zhang", "123");
    subject.login(token);
    //判断是否具有对应角色
    Assert.assertTrue(subject.hasRole("role1"));
    //判断是否具有全部的角色，返回boolean数组，输出结果与传入的角色顺序对应
```

```java
    boolean[] hasRoles = subject.hasRoles(Arrays.asList("role1","role2","role3"));
    System.out.println(hasRoles[0]?"拥有角色role1":"没有role1角色");
    System.out.println(hasRoles[1]?"拥有角色role2":"没有role2角色");
    System.out.println(hasRoles[2]?"拥有角色role3":"没有role3角色");
    //检查单个角色，如果不具有角色，则抛出异常
    subject.checkRole("role1");
    subject.checkRole("role2");
    //检查用户是否具有多个角色，没有则抛出异常
    subject.checkRoles("role1","role3");
}
```

Shiro 中有一个非常重要的角色——Authorizer，此对象用于完成权限验证，前面的案例中并没有使用到该对象。

实际上，授权是在程序运行过程中自动发生的，从源码中可以看到，SecurityManager 对象继承自 Authorizer，如图 6-3 所示。

```java
public interface SecurityManager extends Authenticator, Authorizer, SessionManager {
```

图 6-3 Authorizer 的源码

Authorizer 是 Shiro 进行授权的入口，Shiro 也内置了 Authorizer 实现类。Authorizer 的继承链如图 6-4 所示。

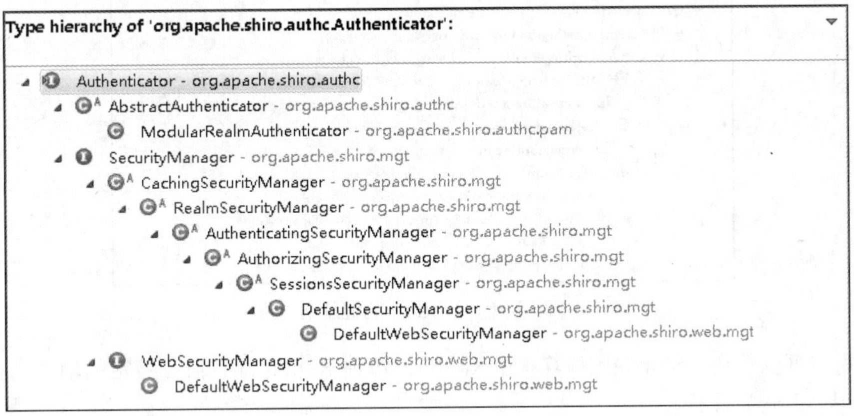

```
Type hierarchy of 'org.apache.shiro.authc.Authenticator':
  Authenticator - org.apache.shiro.authc
    AbstractAuthenticator - org.apache.shiro.authc
      ModularRealmAuthenticator - org.apache.shiro.authc.pam
    SecurityManager - org.apache.shiro.mgt
      CachingSecurityManager - org.apache.shiro.mgt
        RealmSecurityManager - org.apache.shiro.mgt
          AuthenticatingSecurityManager - org.apache.shiro.mgt
            AuthorizingSecurityManager - org.apache.shiro.mgt
              SessionsSecurityManager - org.apache.shiro.mgt
                DefaultSecurityManager - org.apache.shiro.mgt
                  DefaultWebSecurityManager - org.apache.shiro.web.mgt
    WebSecurityManager - org.apache.shiro.web.mgt
      DefaultWebSecurityManager - org.apache.shiro.web.mgt
```

图 6-4 Authorizer 的继承链

Shiro 内置的 Authorizer 实现类为 ModularRealmAuthenticator，根据源码可知此类可以接收多个 Realm，并且验证策略是只要有一个权限验证成功，就返回 true，即用户具有此权限。当调用 Subject 对象的 isPermitted/hasRole 时，验证过程委托给了 SecurityManager 对象完成，而 SecurityManager 对象会将验证继续交由 Authorizer 完成。作为 Shiro 内置的验证对象的 ModularRealmAuthenticator 在验证过程中会进行循环验证，直到返回 true 或 false。验证流程如图 6-5 所示。

```java
public boolean isPermitted(PrincipalCollection principals, String permission) {
    assertRealmsConfigured();
    for (Realm realm : getRealms()) {
        if (!(realm instanceof Authorizer)) continue;
        if (((Authorizer) realm).isPermitted(principals, permission)) {
            return true;
        }
    }
    return false;
}
```

图 6-5 验证流程

当然，这是 Shiro 提供的默认验证规则，用户也可以自定义验证规则，但是在大部分情况下，默认验证规则即可完成用户所需功能。

6.4　Realm

前面对 Realm 进行了简单介绍，简单来说，以前的程序是直接从数据库获取数据，从而进行身份的判断；现在只需要将这件事情委托给 Shiro 完成即可，即告诉 Shiro 数据在哪个地方，Shiro 使用 Realm 就可以完成数据的获取，并进行授权、验证等工作，也可以将 Realm 理解为数据源。

Shiro 主要支持以下几种 Realm。

1．Shiro 内置 Realm

Shiro 内置的 Realm 使用 JDBC 作为数据来源，JDBC 是 Java 连接数据库解决方案。

使用 JDBC 作为数据源的示例在本章 6.2 节中已经介绍过，这里不再重复。下面通过代码深入了解 Realm 的运行机制。

打开 Eclipse，找到 Shiro 提供的 Realm 接口，该接口在 org.apache.shiro.realm 包中，查看 Realm 的继承链，如图 6-6 所示。

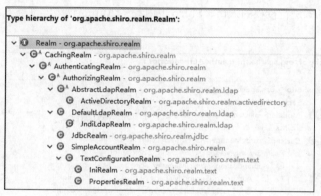

图 6-6　Realm 的继承链

从图 6-6 中可以看到，Shiro 提供了多种 Realm，每种 Realm 都有自己特定的功能，接下来选取部分进行介绍，如表 6-1 所示。

表 6-1　Shiro 内置 Realm 的功能

类名	层级	功能
Realm	1	定义基础的规范，包含 3 个方法： ① getAuthenticationInfo(AuthenticationToken arg0)：用于获取 Token 认证信息，如可以用来验证用户名/密码是否正确； ② public String getName()：获取唯一的 Realm 名称，可以理解成为 Realm 命名； ③ boolean supports(AuthenticationToken arg0)：判断是否支持此 token
CachingRealm	2	提供缓存功能，默认允许缓存，内置了基础缓存，该类为抽象类，通常提供给子类使用。CachingRealm 提供了 getCacheManager()方法，用于获取缓存管理器，需要自行引入。其还实现了 Cache Manager Aware，此接口用于注入 CacheManager 对象。在 Shiro 内部可以使

续表

类名	层级	功能
CachingRealm	2	用 DefaultSecurityManager 组件来检测 Realm 是否实现了 CacheManagerAware 接口，如果是，则可以自动注入相应的 CacheManager 对象，对此有兴趣的读者可以深入查阅源码
AuthenticatingRealm	3	Shiro 提供的验证类，此类为抽象类，主要提供以下方法： ① doGetAuthenticationInfo(AuthenticationToken token)：获取认证信息，为抽象方法，供子类实现； ② getAuthenticationInfo(AuthenticationToken token)：Realm 接口的实现方法，在方法实现中调用了第一个方法； ③ assertCredentialsMatch(AuthenticationToken token, AuthenticationInfo info)：用于校验凭证信息，使用设定的 CredentialsMatcher 对象来完成信息校验
AuthorizingRealm	4	授权 Realm 类，此类为抽象类，主要提供以下方法： ① doGetAuthorizationInfo(PrincipalCollection principals)：授权抽象方法，需要子类实现该方法时才能实现授权； ② getAuthorizationInfo(PrincipalCollection principals)：授权方法，内部调用了第一个方法，所以其功能生效的前提是第一个授权方法被子类实现； ③ AuthorizingRealm 实现了 Authorizer 接口，该接口为授权者，是 Shiro 授权的入口，提供了 checkRole（角色检查）、checkPermission（权限检查）、hasRole（角色判断）、isPermitted（权限判断）等一系列与权限相关的方法；其也实现了 PermissionResolverAware 接口，此接口为权限设置入口；其还实现了 RolePermissionResolverAware 接口，此接口是提供角色权限的设置入口
JdbcRealm	5	通过 JDBC 认证授权，此类为普通类，即开发者可以直接使用的 Realm 类，打开该类，首先看到的是几条 SQL 语句。这些 SQL 语句就是 Realm 获取角色及权限的执行 SQL，如果需要这些 SQL 语句生效，就需要指定 JdbcRealm 的 DataSource，即数据源。此外，数据库创建表、字段时应当遵照 JdbcRealm 定义的表名与字段名。当然，Shiro 允许用户自定义 SQL 语句。JdbcRealm 提供了从数据库获取角色、权限、用户信息的方法
SimpleAccountRealm	5	该类继承自 AuthorizingRealm，意为简单账户域，该类实现了父接口遗留的方法，提供了两个用于存储数据的属性，分别为 Map<String, SimpleRole> roles、Map<String, SimpleAccount> users。但是 SimpleAccount 不关注数据来源，roles 和 users 的数据来源实际上是由 SimpleAccountRealm 的子类 TextConfigurationRealm 获取的
TextConfigurationRealm	6	对文本配置数据提供支持，提供了两个字符串属性 userDefinitions、roleDefinitions，此类主要用于对这两个字符串进行解析，它和其父类一样，没有关注数据来源，而是由子类 IniRealm、PropertiesRealm 完成数据源的获取工作

续表

类名	层级	功能
IniRealm	7	支持 INI 配置文件进行数据源的配置，也是 Shiro 默认提供的 Realm。IniRealm 在 Shiro 中非常重要。在本章第一个示例中，并没有指定 Realm，默认使用 IniRealm 来完成数据源的获取；但是在前面 JdbcRealm 的示例中，在 INI 文件中配置了数据，且 Shiro 获取到了这些数据
PropertiesRealm	7	对 properties 配置文件提供支持，properties 类似于键/值对形式

可以看到，Shiro 提供的 Realm 机制非常简洁，并且十分容易扩展，这为程序开发人员控制程序提供了更好的方式。

当然，此处并没有将 Shiro 中所有的 Realm 都列举出来，读者通过查阅源码可以理解其他 Realm 的机制。

2. 自定义 Realm

Shiro 允许程序开发人员对 Realm 进行扩展，使用自定义 Realm 的前提是当前 Shiro 提供的 Realm 不满足开发需求。如果 Shiro 本身已经提供了某个功能，程序开发人员仍坚持针对该功能编写自定义 Realm，就显得多此一举了。

下面模仿 Shiro 提供的 JdbcRealm 来实现自定义 Realm，以理解自定义 Realm 的实现机制。因为要访问数据库获取数据，所以可直接使用第 1 章中的 DBUtil 类，与该类关联的 Security 类与 druid.properties 配置文件也需要复制到本项目中，数据库的配置信息需要修改为与当前示例一致。

（1）编写 Dao 类，用于获取数据，代码如下。

```java
package com.woniu.dao;

import java.sql.Connection;
import java.sql.PreparedStatement;
import java.sql.ResultSet;
import java.sql.SQLException;
import java.util.HashSet;
import java.util.Set;

import com.woniu.entity.UserRole;
import com.woniu.entity.Users;
import com.woniu.util.DBUtil;

public class UserDao {
    //根据用户名查找用户信息，返回用户对象
    public Users getUsers(Connection conn , String username) throws SQLException {
        String sql = "SELECT * FROM users WHERE username = ?";
        Users users = null;
        PreparedStatement ps = conn.prepareStatement(sql);
        ps.setString(1, username);

        ResultSet rs = ps.executeQuery();
        if(rs.next()) {
            users = new Users();
            users.setId(rs.getInt("id"));
            users.setUsername(rs.getString("username"));
            users.setPassword(rs.getString("password"));
            users.setRoleId(rs.getInt("role_id"));
        }
```

```java
        return users;
    }
    //根据用户名查找角色
    public Set<String> getRoles(Connection conn , String username) throws SQLException {
        String sql = "SELECT rolename FROM users u, user_role r WHERE u.role_id=r.id AND u.username=?";
        Set<String> roles = new HashSet<String>();
        PreparedStatement ps = conn.prepareStatement(sql);
        ps.setString(1, username);
        ResultSet rs = ps.executeQuery();
        if(rs.next()) {
            roles.add(rs.getString("rolename"));
        }
        return roles;
    }
    //根据用户名查找权限
    public Set<String> getPermission(Connection conn , String username) throws SQLException {
        String sql = "SELECT permissionname FROM users u, user_role r , role_permission p WHERE u.role_id=r.id AND p.role_id = r.id AND u.username=?";
        Set<String> roles = new HashSet<String>();
        PreparedStatement ps = conn.prepareStatement(sql);
        ps.setString(1, username);
        ResultSet rs = ps.executeQuery();
        if(rs.next()) {
            roles.add(rs.getString("permissionname"));
        }
        return roles;
    }
    //进行测试
    public static void main(String[] args) throws SQLException {
        UserDao ud = new UserDao();
        Users users = ud.getUsers(DBUtil.getConnection(), "woniu1");
        Set<String> roles = ud.getRoles(DBUtil.getConnection(), "woniu1");
        Set<String> pre = ud.getPermission(DBUtil.getConnection(), "woniu1");
        System.out.println("用户信息:" + users);
        System.out.println("角色信息:" + roles);
        System.out.println("权限信息:" + pre);
    }
}
```

UserDao 类中所用到的 Users、UserRole、RolePermission 类为数据库表的实体映射，相信读者可以自行完成代码编写。UserDao 类的测试结果如图 6-7 所示。

```
Users [id=4, username=woniu1, password=123456, roleId=4]
[system]
[system:*]
```

图 6-7　UserDao 类的测试结果

（2）编写 Realm 类。自定义 Realm 类时，理论上讲，可以选择 Realm 的继承链上任意的类/接口来实现，但是为了方便，并不会直接实现顶级的接口，而是继承有一定功能的某个 Realm，这里选择 AuthorizingRealm，该类已包含 Realm 的基础功能及缓存、验证、授权功能。具体代码如下。

```java
package com.woniu.realm;

import java.sql.Connection;
```

```java
import org.apache.shiro.authc.*;
import org.apache.shiro.authz.AuthorizationInfo;
import org.apache.shiro.authz.SimpleAuthorizationInfo;
import org.apache.shiro.realm.AuthorizingRealm;
import org.apache.shiro.subject.PrincipalCollection;

import com.woniu.dao.UserDao;
import com.woniu.entity.Users;
import com.woniu.util.DBUtil;

public class MyJdbcRealm extends AuthorizingRealm{
    private UserDao userDao = new UserDao();

    //对当前登录成功的用户进行授权
    @Override
    protected AuthorizationInfo doGetAuthorizationInfo(PrincipalCollection principals) {

        String username = (String) principals.getPrimaryPrincipal();
        //获取用户名
        SimpleAuthorizationInfo authorizationInfo = new SimpleAuthorizationInfo();
        Connection conn = null;
        try {
            conn = DBUtil.getConnection();
            authorizationInfo.setRoles(userDao.getRoles(conn,username));
            //设置角色
            authorizationInfo.setStringPermissions(userDao.getPermission(conn,username));  //设置权限
        } catch (Exception e) {
            e.printStackTrace();
        }
        return authorizationInfo;
    }
    //验证方法,用于验证用户信息,即验证用户名与密码是否正确
    @Override
    protected AuthenticationInfo doGetAuthenticationInfo(AuthenticationToken token) throws AuthenticationException {

        String username = (String) token.getPrincipal();   //获取用户名
        Connection conn = null;
        try {
            conn = DBUtil.getConnection();
            Users user = userDao.getUsers(conn, username);
//仅根据用户名查询出的用户信息,不涉及密码
            if (user != null) {
                AuthenticationInfo authcInfo = new SimpleAuthenticationInfo(
                        user.getUsername(), user.getPassword(), "myJdbcRealm");
                return authcInfo;
            } else {
                return null;
            }
        } catch (Exception e) {
            e.printStackTrace();
        }
        return null;
```

        }
    }

（3）编写 INI 配置文件。配置信息仍使用 Shiro 的 INI 配置文件存储，创建 jdbc.ini 文件，添加如下内容。

```
jdbcRealm=com.woniu.realm.MyJdbcRealm
securityManager.realms=$jdbcRealm
```

（4）进行测试，代码如下。

```
@Test
public void test4() {
    IniSecurityManagerFactory factory = new
IniSecurityManagerFactory ("classpath:jdbc.ini");
    SecurityManager sm = factory.getInstance();
    SecurityUtils.setSecurityManager(sm);
    Subject subject = SecurityUtils.getSubject();

    UsernamePasswordToken token = new UsernamePasswordToken("woniu1", "123456");
    try {
        subject.login(token);
        System.out.println("登录成功");
    } catch (Exception e) {
        e.printStackTrace();
        System.out.println("登录失败");
    }
    subject.logout();
}
```

结果显示登录成功。自定义的 JdbcRealm 用于将数据库信息写在另一个配置文件中，而 Shiro 提供的 JdbcRealm 将数据源以 Realm 类属性的方式存储起来，用户使用时需要在 INI 文件中进行指定。

3. 多 Realm

多 Realm 指 Shiro 对 Realm 的支持不是单一的，而是可以支持多个 Realm。Shiro 从 Realm 取出数据进行验证，多个 Realm 的数据如何验证呢？Shiro 提供了 ModularRealmAuthenticator 类以完成授权工作，该类的顶级父接口为 Authenticator，是授权的入口，该类中的 doAuthenticate 是判断 Realm 的方法，其源码如下。

```
protected AuthenticationInfo doAuthenticate(AuthenticationToken authenticationToken)
throws AuthenticationException {
    assertRealmsConfigured();
    Collection<Realm> realms = getRealms();
    if (realms.size() == 1) {
        //单个Realm
        return doSingleRealmAuthentication(
            realms.iterator().next(), authenticationToken);
    } else {
        //多个Realm
        return doMultiRealmAuthentication(
            realms, authenticationToken);
    }
}
```

继续查看源码，在多 Realm 认证中出现了一个对象——AuthenticationStrategy，称为认证策略。Shiro 为多 Realm 提供了以下 3 种认证策略。

（1）AllSuccessfulStrategy：只有当所有的 Realm 都认证成功时才算成功，并且会返回所有 Realm 身份认证成功的认证信息，只要有一个 Realm 认证失败，认证即失败。

（2）FirstSuccessfulStrategy：Realm 中只要有一个认证成功即表示用户认证成功，但是只会返回第一个认证成功的 Realm 的认证信息。

（3）AtLeastOneSuccessfulStrategy：Realm 中只要有一个认证成功即表示用户认证成功，但和 FirstSuccessfulStrategy 有所区别的是，其会返回所有认证成功的 Realm 的认证信息。

在程序开发中，可以根据实际需求，灵活地选择认证策略。

6.5 基于 Shiro 的 Web 开发

V6-1　Shiro 初识

前面主要介绍了 Shiro 的运作机制及基本功能，在实际开发中，Web 程序的开发仍然是主要的，将 Shiro 框架融入 Web 程序是关键。而 Spring 框架通常作为"桥梁"连接各框架，在项目中使用频繁。下面的案例演示了如何将 Shiro 整合到 Spring 管理的 Web 项目中。

1. 创建 Maven 项目并引入依赖

具体代码如下。

```xml
<!-- JUnit单元测试 -->
<dependency>
    <groupId>junit</groupId>
    <artifactId>junit</artifactId>
    <version>4.9</version>
</dependency>
<!-- 日志记录 -->
<dependency>
    <groupId>commons-logging</groupId>
    <artifactId>commons-logging</artifactId>
    <version>1.1.3</version>
</dependency>
<!-- Shiro核心 -->
<dependency>
    <groupId>org.apache.shiro</groupId>
    <artifactId>shiro-core</artifactId>
    <version>1.2.2</version>
</dependency>
<!-- JDBC -->
<dependency>
    <groupId>mysql</groupId>
    <artifactId>mysql-connector-java</artifactId>
    <version>5.1.25</version>
</dependency>
<!-- 数据库连接池 -->
<dependency>
    <groupId>com.alibaba</groupId>
    <artifactId>druid</artifactId>
    <version>1.0.29</version>
</dependency>
<dependency>
    <groupId>javax.servlet</groupId>
    <artifactId>servlet-api</artifactId>
    <version>2.5</version>
    <scope>provided</scope>
</dependency>
<!-- 添加Servlet支持 -->
<dependency>
    <groupId>javax.servlet</groupId>
    <artifactId>javax.servlet-api</artifactId>
```

```xml
            <version>3.1.0</version>
        </dependency>
        <!-- 添加JSP支持 -->
        <dependency>
            <groupId>javax.servlet.jsp</groupId>
            <artifactId>javax.servlet.jsp-api</artifactId>
            <version>2.3.1</version>
        </dependency>
        <!-- 添加JSTL支持 -->
        <dependency>
            <groupId>javax.servlet</groupId>
            <artifactId>jstl</artifactId>
            <version>1.2</version>
        </dependency>
        <dependency>
            <groupId>org.apache.shiro</groupId>
            <artifactId>shiro-web</artifactId>
            <version>1.2.2</version>
        </dependency>
        <dependency>
            <groupId>org.springframework</groupId>
            <artifactId>spring-webmvc</artifactId>
            <version>4.3.8.RELEASE</version>
        </dependency>
        <dependency>
            <groupId>org.springframework</groupId>
            <artifactId>spring-web</artifactId>
            <version>4.3.8.RELEASE</version>
        </dependency>
        <!-- Shiro-Spring集成JAR包 -->
        <dependency>
            <groupId>org.apache.shiro</groupId>
            <artifactId>shiro-spring</artifactId>
            <version>1.2.2</version>
        </dependency>
```

2. 与 Spring 的集成

在 web.xml 文件中配置请求入口，内容如下。

```xml
    <!-- Shiro入口 -->
    <filter>
        <filter-name>shiroFilter</filter-name>
<filter-class>org.springframework.web.filter.DelegatingFilterProxy</filter-class>
        <init-param>
            <!-- 当设置为true时，表示生命周期由ServletContainer管理 -->
            <param-name>targetFilterLifecycle</param-name>
            <param-value>true</param-value>
        </init-param>
    </filter>
    <filter-mapping>
      <filter-name>shiroFilter</filter-name>
      <url-pattern>/*</url-pattern>
    </filter-mapping>

    <!-- 请求入口 -->
  <servlet>
    <servlet-name>dispatcher</servlet-name>
```

```xml
        <servlet-class>org.springframework.web.servlet.DispatcherServlet</servlet-class>
        <init-param>
            <param-name>contextConfigLocation</param-name>
            <param-value>classpath:spring-mvc.xml</param-value>
        </init-param>
    </servlet>
    <!-- 静态资源交由默认Servlet处理 -->
    <servlet-mapping>
        <servlet-name>dispatcher</servlet-name>
        <url-pattern>/</url-pattern>
    </servlet-mapping>
    <servlet-mapping>
        <servlet-name>default</servlet-name>
        <url-pattern>*.jpg</url-pattern>
</servlet-mapping>
<servlet-mapping>
        <servlet-name>default</servlet-name>
        <url-pattern>*.js</url-pattern>
</servlet-mapping>
<servlet-mapping>
        <servlet-name>default</servlet-name>
        <url-pattern>*.css</url-pattern>
</servlet-mapping>
<servlet-mapping>
        <servlet-name>default</servlet-name>
        <url-pattern>*.html</url-pattern>
</servlet-mapping>
```

Spring 的配置如下。

```xml
    <!-- 注解扫描 -->
    <context:component-scan base-package="com.woniu"></context:component-scan>
    <!-- MVC注解支持 -->
    <mvc:annotation-driven/>
    <!-- 静态资源处理 -->
    <mvc:default-servlet-handler/>
    <!-- 视图处理器 -->
    <bean class="org.springframework.web.servlet.view.InternalResourceViewResolver">
        <property name="prefix" value="/jsp/"/>
        <property name="suffix" value=""/>
    </bean>
    <!-- 相当于调用SecurityUtils.setSecurityManager(securityManager) -->
    <bean class="org.springframework.beans.factory.config.MethodInvokingFactoryBean">
        <property name="staticMethod"
                  value="org.apache.shiro.SecurityUtils.setSecurityManager"/>
        <property name="arguments" ref="securityManager"/>
    </bean>
    <!-- 自定义Realm -->
    <bean id="myRealm" class="com.woniu.realm.MyJdbcRealm"/>
    <!-- 安全管理器 -->
    <bean id="securityManager"
          class="org.apache.shiro.web.mgt.DefaultWebSecurityManager">
        <property name="realm" ref="myRealm"/>
    </bean>

    <!-- 基于表单的身份认证过滤器 -->
    <bean id="formAuthenticationFilter"
```

```xml
            class="org.apache.shiro.web.filter.authc.FormAuthenticationFilter">
        <property name="usernameParam" value="username"/>
        <property name="passwordParam" value="password"/>
        <property name="loginUrl" value="/toLogin"/>
    </bean>

    <!-- Shiro的Web过滤器 -->
    <bean id="shiroFilter" class="org.apache.shiro.spring.web.ShiroFilterFactoryBean">
        <!-- Shiro的核心安全接口,此属性是必需的 -->
        <property name="securityManager" ref="securityManager"/>
        <!-- 身份认证失败,跳转到登录页面的配置 -->
        <property name="loginUrl" value="/jsp/login.jsp"/>
        <!-- Shiro连接约束配置,即过滤链的定义 -->
        <property name="filterChainDefinitions">
            <value>
                <!-- 游客身份即可访问 -->
                /login*=anon
                /user/logout*=logout
                <!-- 需要进行身份认证 -->
                /user/index*=authc
                <!-- 需要具有对应角色 -->
                /user/roles=roles[system]
                <!-- 需要拥有相应权限 -->
                /user/list*=perms["user:view"]
            </value>
        </property>
    </bean>
     <bean class="org.springframework.aop.framework.autoproxy.DefaultAdvisorAutoProxyCreator" depends-on="lifecycleBeanPostProcessor"/>
        <bean class="org.apache.shiro.spring.security.interceptor.AuthorizationAttributeSourceAdvisor">
        <property name="securityManager" ref="securityManager"/>
    </bean>
    <!-- Shiro生命周期处理器-->
    <bean id="lifecycleBeanPostProcessor" class="org.apache.shiro.spring.LifecycleBeanPostProcessor"/>
```

Spring 的配置和以前的配置是一样的,但增添了 Shiro 的配置,使用的 Realm 是前面创建的 MyJdbcRealm 类。

Shiro 权限的控制配置在 shiroFilter 中,指定 filterChainDefinitions 为 URL 配置权限。Shiro 的权限配置使用了 ant 类型,支持"*"匹配和"**"匹配,如 user:*表示用于 user 的所有操作权限。

3. Controller 类编写

```java
package com.woniu.controller;

import javax.servlet.http.HttpSession;

import org.apache.shiro.SecurityUtils;
import org.apache.shiro.authc.*;
import org.apache.shiro.subject.Subject;
import org.springframework.stereotype.Controller;
import org.springframework.web.bind.annotation.RequestMapping;
//本类用于处理登录、退出登录等请求
@Controller
@RequestMapping("user")
public class LoginController {
```

```
    @RequestMapping("toLogin")
    public String login() {
        return "login.jsp";
    }
    @RequestMapping("login")
    public String doLogin(String username , String password , HttpSession session) {
        Subject subject = SecurityUtils.getSubject();
        UsernamePasswordToken token = new UsernamePasswordToken(username, password);
        String error = null;
        try {
            subject.login(token);
        } catch (UnknownAccountException e) {
            error = "用户名/密码错误";
        } catch (IncorrectCredentialsException e) {
            error = "用户名/密码错误";
        } catch (AuthenticationException e) {
            error = "其他错误：" + e.getMessage();
        }
        //没有登录成功，跳转到登录页面，并绑定错误信息
        if(error!=null) {
            session.setAttribute("error", error);
            return "login.jsp";
        }
        return "redirect:index";
    }
    //退出登录
    @RequestMapping("logout")
    public String logout(HttpSession session) {
        session.invalidate();
        return "login.jsp";
    }
    //需要经过身份认证才能进行访问
    @RequestMapping("index")
    public String index() {
        return "index.jsp";
    }
    //需要有user:view权限
    @RequestMapping("list")
    public String list() {
        return "index.jsp";
    }
    //需要是system角色才能访问
    @RequestMapping("roles")
    public String roles() {
        return "index.jsp";
    }
}
```

4. JSP 页面编写

（1）编写登录页面 login.jsp，代码如下。

```
<%@ page contentType="text/html;charset=UTF-8" language="java" %>
<html>
<head>
    <title>登录</title>
    <style>.error{color:red;}</style>
</head>
```

```
<body>
<div class="error">${error}</div>
<form action="${pageContext.request.contextPath}/user/login" method="post">
    用户名：<input type="text" name="username"><br/>
    密码：<input type="password" name="password"><br/>
    <input type="submit" value="登录">
</form>
</body>
</html>
```

（2）编写登录成功页面 index.jsp，代码如下。

```
<%@ page contentType="text/html;charset=UTF-8" language="java" %>
<html>
<head>
    <title>首页</title>
</head>
<body>
登录成功！<a href="${pageContext.request.contextPath}/user/login">退出</a>
</body>
</html>
```

5．结果测试

（1）身份认证测试。在登录之前访问 http://localhost:8080/shiro-web/user/index 请求，因为此请求要求经过身份认证才能访问，所以 Shiro 会将请求页面重定向到 shiroFilter 中配置的 loginUrl 路径。使用任意账户进行登录，登录成功之后跳转到 index.jsp。此时，使用当前浏览器打开新页面，再次访问 http://localhost:8080/shiro-web/user/index，可以访问此页面，因为用户已经登录，也就具有了访问 index 请求的权限。

（2）角色认证测试。roles 请求路径要求只有 system 角色才能访问。使用 woniu2 账户进行登录，登录之后访问 http://localhost:8080/shiro-web/user/roles，发现该页面无法正常访问，而是返回了 401 错误，表示账户无访问权限，如图 6-8 所示。

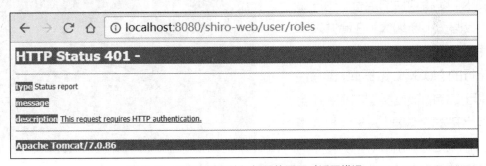

图 6-8　账户访问无访问权限的页面时返回错误

使用 woniu1 账户登录，并访问 roles 请求，此时，页面正常显示，证明角色访问限制配置成功。

（3）权限认证测试。使用 woniu3 账户登录并访问 list 请求，仍无法正常访问，使用 woniu2 账户登录并访问 list 请求，页面正常显示，权限认证配置成功。在程序开发中，可以通过对 URL 的配置完成权限的精确控制。

通过本章的学习，读者应掌握权限控制的基本使用，权限对项目的安全有着非常重要的作用，无论是复杂的大型项目，还是简单的小型项目，Shiro 都有可用之处。

V6-2　Shiro 应用

第7章

Redis

本章导读

■ 使用 SSM 及 Shiro，已经能够满足很多项目的要求。随着项目的运行，用户量不断增加、数据量不断增大，在高并发下提高访问速度是系统面临的问题。Redis 作为当下流行的 NoSQL 数据库，读写性能极高，并且支持的数据类型丰富，完全能够满足程序开发的需要。对于大数据量的操作，使用 NoSQL 是很好的办法。通过 Redis，可以极大地提高程序的性能。本章将深入探讨 Redis 在程序开发中的使用。

学习目标

（1）了解Redis的由来。
（2）掌握Redis的基本操作。
（3）掌握Jedis对Redis的访问。

7.1 认识 Redis

7.1.1 RDBMS 与 NoSQL

关系数据库管理系统（Relational Database Management System，RDBMS）是目前运用最多、使用范围最广的数据库系统，其发展历史悠久，可追溯到 20 世纪 60 年代，读者耳熟能详的 Oracle、MySQL、SQL Server 等都属于关系型数据库。关系型数据库已经深入到各行各业中，如银行、金融、传统企业、互联网企业，甚至日常生活都离不开关系型数据库，浏览器、App 甚至家庭电视都与关系型数据库相关。

数据管理由来已久，网状结构数据库出现在 20 世纪 60 年代初，美国通用电气公司的员工巴赫曼（Bachman）等成功研发出世界上第一个网状结构数据库——IDS。网状结构数据库在当时得到了广泛运用，而层次结构数据库出现在 20 世纪 60 年代末，比较出名的是 IBM 公司出品的 IMS。

虽然网状结构数据库和层次结构数据库可以解决数据的集中和共享等问题，但是其对数据独立性和抽象性问题没有很好的解决办法。1970 年，埃德加·弗兰克·科德（Edgar Frank Codd）发表了为关系型数据库奠定基础的一篇论文——"A Relational Model of Data for Large Shared Data Banks"。1974 年，雷·博伊斯（Ray Boyce）和唐·钱伯林（Don Chamberlin）提出了结构化查询语言（Structured Query Language，SQL），现在使用的 SQL 是已经经过美国国家标准协会多次推行修订之后的版本。

关系型数据库理论的提出，让很多企业看到了巨大的商业市场，甲骨文公司的数据库产品 Oracle 是最为成功的关系型数据库之一，而著名的数据库产品 MySQL 目前也属于甲骨文公司。

RDBMS 的数据存储在被称为表的对象中，而表中的数据结构以行和列的形式进行展现，对数据的查询操作也有相关的数据理论作为基础，常用的并、差等代数理论在数据库中都有体现。

随着互联网的发展，用户量、数据量不断增加，传统关系型数据库的瓶颈也越来越明显。试想，当在网络中访问某个数据的延迟有几秒的时候，用户的体验会是多么糟糕，所以数据的处理速度是每个程序都应该考虑的问题。关系型数据库的瓶颈就是 I/O 的瓶颈，因为当用户并发量比较高的时候，甚至可以达到每秒上万次读写请求，磁盘的 I/O 瓶颈也就变成了数据库的瓶颈。解决这个问题比较传统的方案是采用数据库分区分表，即对数据量大的表进行拆分，也可以使用数据库的主从机制实现读写分离。

而关系型数据库的缺点恰好是非关系型数据库的优点，非关系型数据库对大数据量、高并发处理相当擅长，其被称为 NoSQL，即 Not Only SQL。目前 NoSQL 主要有以下几种类型。

1．列式存储

其代表为 HBase。HBase 是 Apache 顶级开源项目 Hadoop 的子项目，是基于 Google 大数据处理 BigTable 的开源版本，采用 Java 语言编写，并且可以搭配 Hadoop 完善 Hadoop 生态系统。

2．文档存储

其代表为 MongoDB。文档存储，顾名思义，指将数据存储在文档之中。MongoDB 的数据格式类似于 JSON，而 JSON 的扩展性非常强，这也对数据库的扩展起到了很好的效果。

3．键/值对存储

其代表为 Memcached、Redis。键/值对即 Key-Value 形式，这种类型的数据库通常不限定 Value 的类型，而 Redis 对此做了限定，这些内容会在后面章节中进行详细介绍。

4．图存储

其代表为 Neo4j，它表示生活中真正的关系梳理，是图形关系的最好表现。例如，在人脉关系中，张三喜欢李四，李四喜欢王五，王五喜欢张三，这样的关系使用关系型数据库表示会非常困难，即便表示了关系纽带，在进行查询时效率也很低。而图存储恰好解决了这类问题。

以上对 NoSQL 进行了初步的介绍，而 Redis 作为 NoSQL 的一种，深受程序开发人员的喜爱，Redis 是以内存运行并可支持持久化的一种 NoSQL。众所周知，内存运行的速度是远远高于磁盘的，相对于传统关系型数据库的磁盘运行速度，NoSQL 的速度明显快很多。

7.1.2 Redis 安装

Redis 非常小巧，并且支持 Windows 和 Linux 操作系统。

1．在 Windows 中安装 Redis

（1）下载 Redis。Redis 在其官网和 GitHub 上都可以下载，下载完成之后直接将其解压到指定目录中即可，这里解压到 D:\安装\Redis 目录中。

（2）启动 Redis。Redis 是免安装的，即可以直接运行使用。启动 Redis 时需要指定 Redis 配置文件，该文件就是 Redis 安装目录中的 redis.windows.conf，该文件提供了很多默认配置，已经足够用户使用。打开命令行窗口，进入 Redis 目录，输入启动命令即可启动 Redis，启动命令如图 7-1 所示。

图 7-1　启动 Redis

弹出图 7-2 所示的信息时表示 Redis 启动成功。

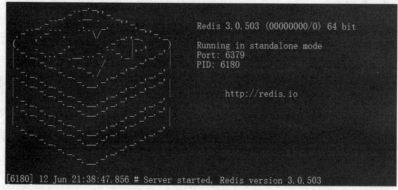

图 7-2　Redis 启动成功

2．客户端使用

Redis 启动完成后，切记不要关闭启动的命令行窗口。再打开一个新的命令行窗口，进入 Redis 安装目录，输入客户端启动命令，启动客户端，如图 7-3 所示。

图 7-3　启动客户端

输入命令之后，若显示 IP 地址：端口号，则表示客户端启动成功。

3．测试

Redis 是键/值对形式的数据库，操作命令也非常简单，这里只进行简单测试。Redis 的命令如表 7-1 所示。

表 7-1　Redis 的命令

命令	功能
set key value	存储一个键/值对到 Redis 数据库中
get key	获取 key 对应的 value

测试效果如图 7-4 所示。

```
127.0.0.1:6379> set name zhangsan
OK
127.0.0.1:6379> get name
"zhangsan"
127.0.0.1:6379>
```

图 7-4　测试效果

4. 密码登录

在前面的测试中，可以发现 Redis 客户端的操作是不需要密码的，这在测试环境中不会出现问题，但是在开发环境中，这样是极不安全的，所以需要对 Redis 进行密码登录配置。

Redis 中并没有用户系统，只包含了一个轻量级的 auth 账户，使用密码认证即可。打开 Redis 根目录中的 redis.windows-service.conf，找到 requirepass foobared 行，去掉前面的注释，foobared 就是需要验证的密码，为了方便记忆，将其改为自己熟悉的密码即可。

需要注意的是，去掉注释时，此行前面的空格也要一并去掉，否则 Redis 会启动失败。修改完成后重启 Redis（关闭命令行窗口，重新启动 Redis），重新连接客户端（输入 exit 命令退出，并重新连接），再次输入 set 命令，此时并不能设置成功，而是出现需要认证的提示信息，如图 7-5 所示。

```
D:\安装\Redis>redis-cli.exe
127.0.0.1:6379> set name lisi
(error) NOAUTH Authentication required.
127.0.0.1:6379> _
```

图 7-5　需要认证的提示信息

输入命令"auth password"，password 是前面设置的密码，弹出"OK"提示表示认证成功，再次运行"set"命令即可成功存储。

5. Windows 服务开启

前面 Redis 的运行并不是以 Windows 服务的形式存在的，命令行窗口必须一直打开，关闭命令行窗口就表示关闭了 Redis。可以将 Redis 设置为 Windows 的服务，从而省去这些麻烦。

打开命令行窗口，进入 Redis 根目录，输入命令"redis-server --service-install redis.windows.conf --loglevel verbose"，Windows 会提示修改，选择"是"即可。如果没有出现错误提示，则表示服务设置成功，此时打开 Windows 服务系统，可以看到增加了 Redis 服务，如图 7-6 所示。

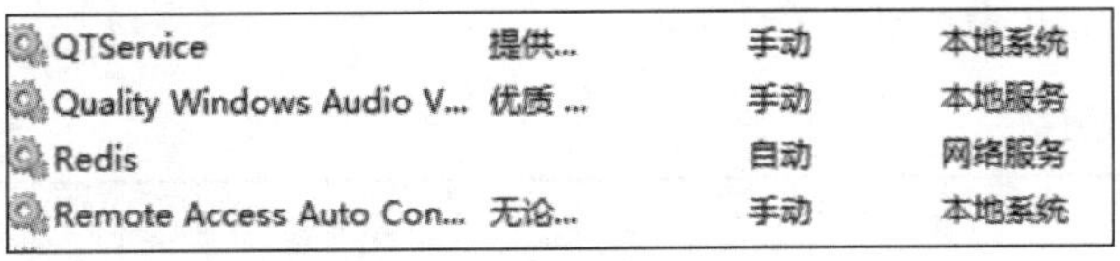

图 7-6　增加了 Redis 服务

可以直接右键单击该服务进行启动、关闭、卸载等操作，也可以使用命令进行操作。Redis 服务的命令如表 7-2 所示。

表 7-2 Redis 服务的命令

命令	功能
redis-server --service-install redis.windows.conf --loglevel verbose	将 Redis 设置为服务
redis-server --service-start	启动服务
redis-server --service-stop	关闭服务
redis-server --service-uninstall	卸载服务

6. 在 Linux 中安装 Redis

Linux 的版本众多，这里选择企业开发使用较多的 Ubuntu 版本安装 Redis。

运行命令":sudo apt-get install redis-server"，系统会自动安装 Redis。如果出现找不到 redis-server 包错误，可先运行"sudo apt-get update"命令更新列表。

安装成功之后 Redis 会自动启动，运行"ps -aux|grep redis"命令，查看 Redis 能否运行。Redis 成功运行的效果如图 7-7 所示。

图 7-7 Redis 成功运行的效果

运行命令"redis-cli"，启动 Redis 客户端，如图 7-8 所示。启动之后的效果和 Windows 是一样的，此后即可进行操作。

图 7-8 启动 Redis 客户端

7.1.3 Redis 命令

学习 Redis 命令之前，需要了解 Redis 所支持的数据类型。Redis 支持的数据类型相对于其他 NoSQL 数据库而言丰富得多，例如，Memcached 仅支持字符串（String）类型，而 Redis 所支持的数据类型包括字符串（String）、哈希（Hash）、列表（List）、集合（Set）和有序集合（Sorted Set）。Redis 支持的数据类型也是程序开发中经常用到的类型，相信读者是非常熟悉的。Redis 对不同数据类型的操作命令也不同。

1. String

前面已经使用了 String 类型的两个命令——set 和 get，分别用于设置 key-value 和获取 key 对应的 value。String 类型的常用命令如表 7-3 所示。

表 7-3 String 类型的常用命令

命令	功能
getrange key start end	获取 key 的 value，并截取从 start 到 end 下标的子串，在 Redis 中，字符串下标从 0 开始，到字符串长度-1 结束，这一点和 Java 是一样的
getset key value	为 key 设置值，如果该 key 原本有值，则返回旧值，若没有值，则返回 nil，不管有没有旧值，都不会影响新值的设定
strlen key	获取 key 存储字符串的长度

续表

命令	功能
append key value	追加 value 到 key 的旧值的后面,如果不存在旧值,则表示设置 key-value
setnx key value	避免 key 覆盖的命令,仅当 key 不存在时设置 key-value,而 set 命令无论 key 是否存在都会进行设置
setrange key offset value	将 key 对应的 value 从下标 offset 开始替换为 value,如果 key 不存在,则下标前面将会使用空格替代
setex key seconds value	设置 key-value,且 key 的有效时间为 seconds,以秒为单位

此处并没有罗列出 String 类型的所有命令,仅对其常用命令进行了说明,如果需要使用更多的命令,可以到 Redis 官网查询。

String 类型常用命令的测试结果如图 7-9 所示。

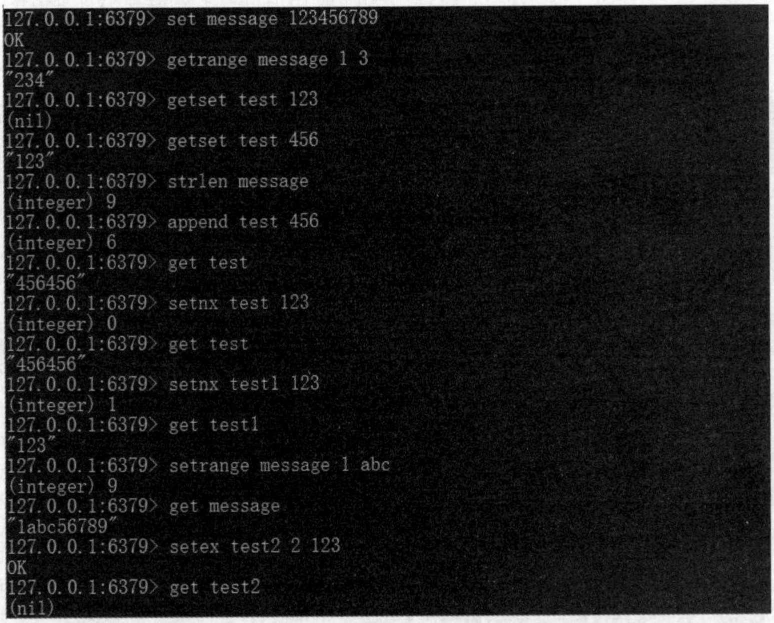

```
127.0.0.1:6379> set message 123456789
OK
127.0.0.1:6379> getrange message 1 3
"234"
127.0.0.1:6379> getset test 123
(nil)
127.0.0.1:6379> getset test 456
"123"
127.0.0.1:6379> strlen message
(integer) 9
127.0.0.1:6379> append test 456
(integer) 6
127.0.0.1:6379> get test
"456456"
127.0.0.1:6379> setnx test 123
(integer) 0
127.0.0.1:6379> get test
"456456"
127.0.0.1:6379> setnx test1 123
(integer) 1
127.0.0.1:6379> get test1
"123"
127.0.0.1:6379> setrange message 1 abc
(integer) 9
127.0.0.1:6379> get message
"1abc56789"
127.0.0.1:6379> setex test2 2 123
OK
127.0.0.1:6379> get test2
(nil)
```

图 7-9　String 类型常用命令的测试结果

2. Hash

Hash 的数据结构是非常适合存储对象的,Hash 本身也是键/值对的形式,而对于 Redis 来说,Hash 的名称与 Hash 的值是一个键/值对。Hash 类型的常用命令如表 7-4 所示。

表 7-4　Hash 类型的常用命令

命令	功能
hmset person name "zhang san" age 25	为 Hash 表中的 person 添加两个属性——name 和 age,此命令可用于设置多个属性
hset person sex "M"	为 Hash 表中的 person 设置属性 sex 的值,该命令仅能设置一个属性
hexists person sex	判断 Hash 表中 person 的 sex 属性是否存在,返回 int 型数据,0 表示不存在,1 表示存在

续表

命令	功能
hget person name	获取 Hash 表中 person 的 name 属性的值，仅能获取一个属性的值
hmget person name age	获取 Hash 表中 person 的 name、age 属性的值，该命令用于获取一个或多个属性的值
hgetall person	获取 Hash 表中 person 的所有 key 和 value，要慎用，因为 Redis 的每一个 Hash 表可以存储 $2^{32}-1$ 个键/值对，有可能返回比较庞大的数据量
hkeys person	获取 Hash 表中 person 的所有 key
hlen person	获取 Hash 表中 person 的所有 key 的数量
hdel person name	删除 Hash 表中 person 的 name 属性，支持同时删除多个属性，删除成功返回 1，删除失败返回 0

Hash 类型常用命令的测试结果如图 7-10 所示。

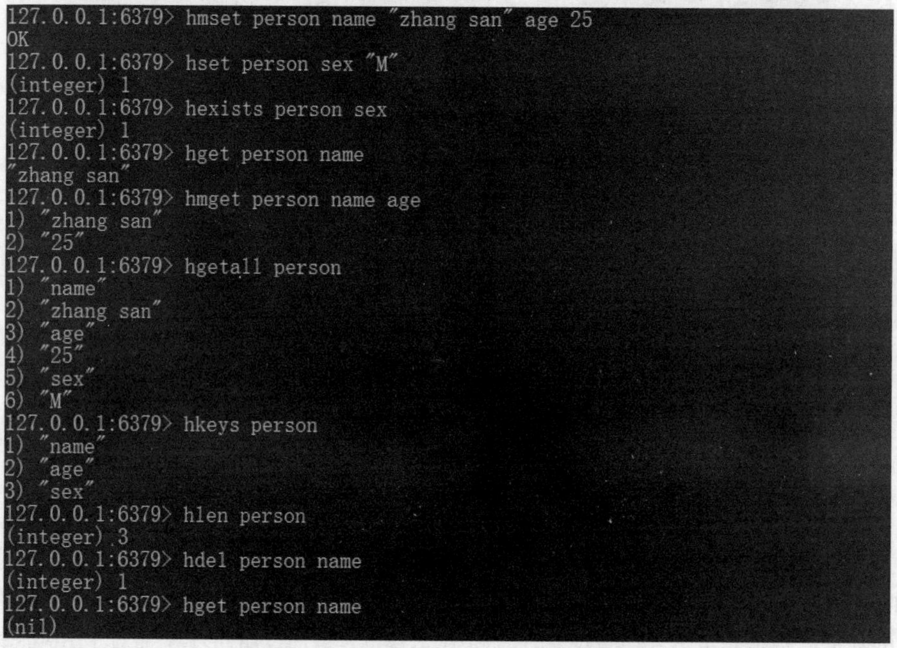

```
127.0.0.1:6379> hmset person name "zhang san" age 25
OK
127.0.0.1:6379> hset person sex "M"
(integer) 1
127.0.0.1:6379> hexists person sex
(integer) 1
127.0.0.1:6379> hget person name
"zhang san"
127.0.0.1:6379> hmget person name age
1) "zhang san"
2) "25"
127.0.0.1:6379> hgetall person
1) "name"
2) "zhang san"
3) "age"
4) "25"
5) "sex"
6) "M"
127.0.0.1:6379> hkeys person
1) "name"
2) "age"
3) "sex"
127.0.0.1:6379> hlen person
(integer) 3
127.0.0.1:6379> hdel person name
(integer) 1
127.0.0.1:6379> hget person name
(nil)
```

图 7-10　Hash 类型常用命令的测试结果

3. List

Redis 支持列表数据类型，列表数据类型是顺序的数据结构，使用下标表示列表元素的位置。List 类型的常用命令如表 7-5 所示。

表 7-5　List 类型的常用命令

命令	功能
lpush key value1 [value2...]	向列表头部插入一个或多个列表元素，列表不存在时会新建一个列表

续表

命令	功能
lpop key	获取列表头部的第一个元素,并且会移除该元素
lindex key index	通过下标获取列表元素
lpushx key value	向列表插入一个元素,该列表必须存在,否则插入失败
lrange key start end	获取从 start 到 end 下标内的元素,List 的下标从 0 开始,到长度-1 结束
lset key index value	为指定下标的元素设置值,不能超越下标进行设置
ltrim key start end	将列表指定下标范围以外的元素移除
rpop key	获取列表的最后一个元素,并且移除该元素
rpushx key value	为列表设置一个值,且只能设置一个值,该列表必须存在,不存在时将会导致设置失败
rpush key value1 [value2]	为列表设置一个到多个值,列表不存在时会新建一个列表
llen key	获取列表的长度

List 类型常用命令的测试结果如图 7-11 所示。

```
127.0.0.1:6379> lpush mylist a b c d e f
(integer) 6
127.0.0.1:6379> lpop mylist
"f"
127.0.0.1:6379> lpushx list g
(integer) 0
127.0.0.1:6379> lpushx mylist g
(integer) 6
127.0.0.1:6379> lrange mylist 1 3
1) "e"
2) "d"
3) "c"
127.0.0.1:6379> lset mylist 3 x
OK
127.0.0.1:6379> ltrim mylist 1 3
OK
127.0.0.1:6379> rpop mylist
"x"
127.0.0.1:6379> rpushx mylist y
(integer) 3
127.0.0.1:6379> rpush mylist l m n
(integer) 6
127.0.0.1:6379> llen mylist
(integer) 6
```

图 7-11 List 类型常用命令的测试结果

4. Set

Redis 中的 Set 类型称为集,是无序不可重复的数据的集合,Set 中存储的数据是 String 类型的。Set 类型的常用命令如表 7-6 所示。

表 7-6 Set 类型的常用命令

命令	功能
sadd key value1 [value2 ...]	增加一个或多个值到集合中,如果集合不存在,则会新建一个集合
scard key	获取集合中元素的个数
sismember key value	判断集合是否包含指定的 value 值,返回值为 1 表示包含,返回值为 0 表示不包含

续表

命令	功能
smembers key	返回集合中的所有元素
spop key	将集合中的一个随机元素移除,并返回该元素
srem key value1 [value2...]	将指定的元素从集合中移除,支持多元素同时移除
sscan key index [match pattern] [count count]	从指定的位置循环迭代集合中的元素,pattern 为正则表达式,可选;count 是最多迭代的个数,可选
srandmember key [count]	随机从集合中返回指定数量的元素,默认为 1

Set 类型常用命令的测试结果如图 7-12 所示。

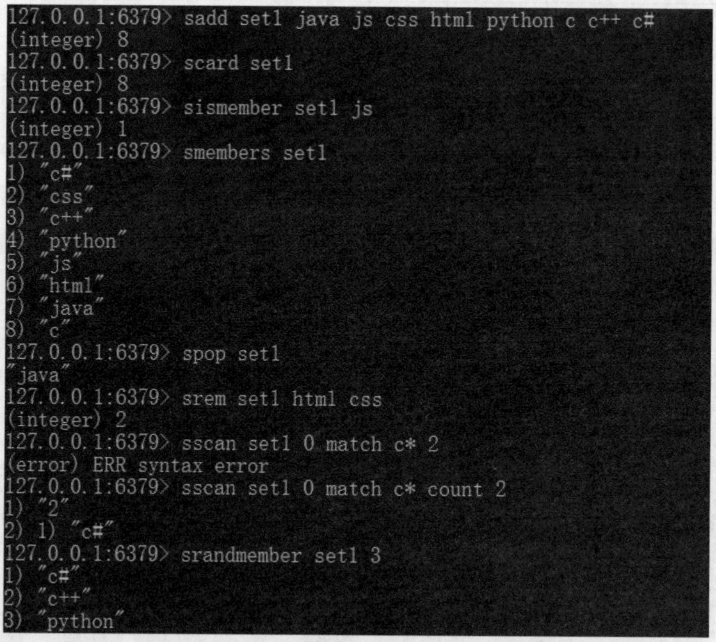

图 7-12 Set 类型常用命令的测试结果

5. Sorted Set

Sorted Set 是 Redis 中的有序集合,有序集合不能包含重复的数据。有序集合是在集合的基础之上,为每个元素增加一个与之关联的分数。分数是 double 类型,用于进行排序。分数和元素是需要区分开的,元素不能重复,而分数是可以重复的,当分数相同时,Redis 会根据字典顺序来进行排序。Sorted Set 的常用命令如表 7-7 所示。

表 7-7 Sorted Set 的常用命令

命令	功能
zadd key score1 value1 [score2 value2]	向 Sorted Set 中添加一个到多个元素。如果该集合不存在,则会创建一个集合;如果指定的元素存在,则表示更新此元素的分数
zcard key	返回 Sorted Set 的元素个数
zcount key min max	返回在指定分数 min~max 范围内的元素数量
zrank key value	返回 value 元素在 Sorted Set 中的索引

续表

命令	功能
zrem key value1 [value2...]	将 Sorted Set 中的一个或多个元素移除，移除成功返回 1，失败返回 0
zscore key value	返回 Sorted Set 中的成员分数值
zrevrank key value	返回 Sorted Set 中指定成员的排名，在 Sorted Set 中排名是根据分数值从大到小排序的
zincrby key increment value	将 Sorted Set 中指定元素的分数在当前基础之上加 increment
zrange key start end	返回 Sorted Set 指定下标内的元素，如果是 0~1 区间，则表示返回所有元素

Sorted Set 类型常用命令的测试结果如图 7-13 所示。

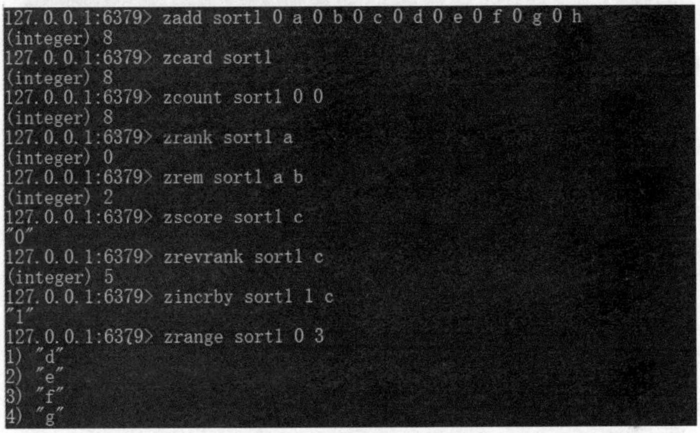

```
127.0.0.1:6379> zadd sort1 0 a 0 b 0 c 0 d 0 e 0 f 0 g 0 h
(integer) 8
127.0.0.1:6379> zcard sort1
(integer) 8
127.0.0.1:6379> zcount sort1 0 0
(integer) 8
127.0.0.1:6379> zrank sort1 a
(integer) 0
127.0.0.1:6379> zrem sort1 a b
(integer) 2
127.0.0.1:6379> zscore sort1 c
"0"
127.0.0.1:6379> zrevrank sort1 c
(integer) 5
127.0.0.1:6379> zincrby sort1 1 c
"1"
127.0.0.1:6379> zrange sort1 0 3
1) "d"
2) "e"
3) "f"
4) "g"
```

图 7-13 Sorted Set 类型常用命令的测试结果

6. 键操作

前面的操作都是针对特定的类型进行的，而 Redis 还提供了对所有键通用的操作命令，如删除、判断等常用操作命令。键操作的常用命令如表 7-8 所示。

表 7-8 键操作的常用命令

命令	功能
del key	删除 key，删除成功返回 1，失败返回 0
exists key	判断 key 是否存在，存在返回 1，不存在返回 0
expire key seconds	设置 key 的过期时间，单位为秒，设置成功返回 1，失败返回 0
persist key	将 key 的过期时间移除，移除成功返回 1，失败返回 0
pttl key	返回 key 的过期时间值，单位为毫秒
keys pattern	返回匹配正则 pattern 的 key
randomkey	从已经存在的 key 中随机返回一个
rename key newkey	修改 key 的名称
type key	返回 key 的数据类型

键操作常用命令的测试结果如图 7-14 所示。

```
127.0.0.1:6379> set k1 1
OK
127.0.0.1:6379> set k2 2
OK
127.0.0.1:6379> set ke3 3
OK
127.0.0.1:6379> set k4 4
OK
127.0.0.1:6379> del k1
(integer) 1
127.0.0.1:6379> exists k1
(integer) 0
127.0.0.1:6379> expire k2 10
(integer) 1
127.0.0.1:6379> persist k2
(integer) 1
127.0.0.1:6379> expire k2 50
(integer) 1
127.0.0.1:6379> pttl k2
(integer) 44484
127.0.0.1:6379> keys k*
1) "ke3"
2) "k4"
3) "k2"
127.0.0.1:6379> randomkey
"set1"
127.0.0.1:6379> rename set1 k5
OK
127.0.0.1:6379> type set1
none
```

图 7-14　键操作常用命令的测试结果

7.2 Jedis 访问 Redis

7.2.1 常用 API

在系统开发时常用到 JDBC 技术，它是 Java 程序访问数据库的解决方案，它提供了一套标准接口，由数据库厂商提供标准接口的实现，从而基于接口进行编程，使任何 Java 程序开发人员都能使用同样的方式访问不同的数据库。

但对 NoSQL，Java 并没有提供像 JDBC 这样的接口，要使用 Java 访问 NoSQL 时，需要使用每个 NoSQL 厂商提供的访问特定数据库的 JAR 包，与 JDBC 不同的是，这些 JAR 包的操作 API 都不尽相同。要使用 Java 访问 Redis，就需要使用 Redis 提供的 JAR 包，称为 Jedis。使用 Jedis 需要 Jedis 包的支持。可以直接下载 JAR 包，也可以使用 Maven 将 Jedis 依赖导入到项目中。

V7-1　Redis 客户端使用

1. 在项目中导入 Jedis 支持

创建 Maven 项目 Redis，导入 Jedis 支持，其依赖如下。

```
<!-- Jedis -->
<dependency>
    <groupId>redis.clients</groupId>
    <artifactId>jedis</artifactId>
    <version>2.9.0</version>
</dependency>
```

2. 创建 RedisTest 类并进行测试

具体代码如下。

```
@Test
public void test1() {
    //连接Redis服务，localhost表示本机
    Jedis jedis = new Jedis("localhost");
```

```
        //测试连接,连接正常会返回pong字符串
        String ping = jedis.ping();
        System.out.println(ping);
        //关闭连接,和JDBC一样,连接是非常消耗资源的
        //无法自动关闭连接,需要调用close方法将连接关闭
        jedis.close();
}
```

想象中的运行结果并没有出现,而是抛出了异常,即 Jedis 连接数据库异常,如图 7-15 所示。

```
redis.clients.jedis.exceptions.JedisDataException: NOAUTH Authentication required.
    at redis.clients.jedis.Protocol.processError(Protocol.java:127)
    at redis.clients.jedis.Protocol.process(Protocol.java:161)
    at redis.clients.jedis.Protocol.read(Protocol.java:215)
    at redis.clients.jedis.Connection.readProtocolWithCheckingBroken(Connection.java:340)
    at redis.clients.jedis.Connection.getStatusCodeReply(Connection.java:239)
```

图 7-15　Jedis 连接数据库异常

这是认证失败的异常。回顾之前进行的配置,访问 Redis 时设置了 auth 密码认证,而使用 Jedis 访问 Redis 时,同样需要进行相应设置。在创建 Jedis 对象之后添加代码,进行认证设置,代码如下。

```
//进行auth密码认证
jedis.auth("123456");
```

保证密码正确,即可认证成功。

3. 连接池设置

连接资源十分重要,而频繁地创建及销毁连接是对资源的浪费,Jedis 提供了 JedisPool 来封装 commons-pool,以进行连接池管理,创建连接池代码如下。

```
@Test
public void test2() {
    //创建Jedis连接池
    JedisPool pool = new JedisPool("localhost");
    //通过连接池获取Jedis连接
    Jedis jedis = pool.getResource();
    jedis.auth("123456");
    String ping = jedis.ping();
    System.out.println(ping);
    //销毁连接池
    pool.destroy();
}
```

4. 连接池配置

Redis 提供了 JedisPoolConfig 类来支持连接池配置,操作代码如下。

```
@Test
public void test3() {
    JedisPoolConfig config = new JedisPoolConfig();
    /*当新任务到达而没有空闲的连接时,是否阻塞任务
    直到超时,默认值为true,设置为false时会直接抛
    出异常,阻止任务的提交*/
    config.setBlockWhenExhausted(false);
    /*设置Pool的最大空闲连接数,默认值为8,当空
    闲连接数超过此设定值时,Pool会销毁多余的连接*/
    config.setMaxIdle(5);
    //设置Pool的最大连接数,默认值为8
    config.setMaxTotal(10);
    /*设置任务阻塞最大等待时间(单位为毫秒),当setBlockWhenExhausted
```

```
        设置为false时，此设置无效*/
        config.setMaxWaitMillis(5000);
        //设置最小空闲连接数，默认为0
        config.setMinIdle(2);
        //获取Jedis连接时检测此连接是否有效，默认不检测
        config.setTestOnBorrow(true);
        JedisPool pool = new JedisPool("localhost");
        Jedis jedis = pool.getResource();
        jedis.auth("123456");
        String ping = jedis.ping();
        System.out.println(ping);
        pool.destroy();
    }
```

5. Jedis 基本操作

Redis 对不同数据类型的操作命令不同，同样，Jedis 对不同数据类型的 API 也不尽相同。为了方便测试，将获取的 Jedis 对象封装为一个方法。

（1）String 操作，代码如下。

```
@Test
public void test() throws InterruptedException {
    Jedis jedis = getJedis();
    //清空Redis中的所有键/值对，测试时可以使用
    jedis.flushDB();
    System.out.println("新增键/值对~~~~~~~~~~~~~~~~~~~~~~~~~~~~~~~~~~~");
    //新增键/值对，设置成功返回OK
    System.out.println(jedis.set("k1","value1"));
    System.out.println(jedis.set("k2","value2"));
    System.out.println(jedis.set("k3", "value3"));
    //删除成功返回OK，删除失败(如没有该key)返回0
    System.out.println("删除k2:"+jedis.del("k2"));
    //获取key的值，key不存在时返回null
    System.out.println("获取k2的值:"+jedis.get("k2"));
    //在key的value最后追加字符串，如果key不存在，则新建一个key
    System.out.println("追加字符串:"+jedis.append("k3", "End"));
    System.out.println("k3的值: "+jedis.get("k3"));
    //mset，设置成功返回OK
    System.out.println("同时设置多个键/值对："+jedis.mset("k1","v1","k2","v2","k3","v3"));
    //mget，以字符串数组的形式返回，key不存在时返回null
    System.out.println("获取多个键/值对的值："+jedis.mget("k1","k2","k3"));
    System.out.println("同时获取多个key的值："+jedis.mget("k1","k2","k3","k4"));
    System.out.println("同时删除多个key的值："+
    jedis.del(new String[]{"k1","k2"}));
    System.out.println("同时获取多个key的值："+jedis.mget("key01","key02","key03"));
    jedis.flushDB();
    //setnx，在key不存在的情况下设置才会成功
    System.out.println(jedis.setnx("k1", "value1"));
    System.out.println(jedis.setnx("k2", "value2"));
    System.out.println(jedis.setnx("k2", "value2-new"));
    System.out.println(jedis.get("k1"));
    System.out.println(jedis.get("k2"));
```

```java
        //设置键/值对，并设置有效时间，单位为s
        System.out.println(jedis.setex("k3", 2, "value3"));
        System.out.println(jedis.get("k3"));
        //测试key失效，阻塞3s
        TimeUnit.SECONDS.sleep(3);
        System.out.println(jedis.get("k3"));
        //获取key的值并为其设置新的值
        System.out.println(jedis.getSet("k2", "k2GetSet"));
        System.out.println(jedis.get("k2"));
        /*getrange,获取下标2~4的子串,包含下标2但不包含下标4,下标从0开始,到长度-1结束*/
        System.out.println("获取k2的值的子串："+
                    jedis.getrange("key2", 2, 4));
}
```

String 操作测试结果如图 7-16 所示。

```
新增键/值对~~~~~~~~~~~~~~~~~~~~~~
OK
OK
OK
删除k2:1
获取k2的值:null
追加字符串:9
k3的值：value3End
同时设置多个键/值对：OK
获取多个键/值对的值：[v1, v2, v3]
同时获取多个key的值：[v1, v2, v3, null]
同时删除多个key的值：2
同时获取多个key的值：[null, null, null]
1
1
0
value1
value2
OK
value3
null
value2
k2GetSet
获取k2的值的子串：
```

图 7-16　String 操作测试结果

（2）List 操作，代码如下。

```java
@Test
    public void test4() {
        Jedis jedis = getJedis();
        jedis.flushDB();
        //向列表中添加元素,支持添加一个或多个元素
        jedis.lpush("myList", "List", "Set", "Map", "Tree");
        jedis.lpush("myList", "ArrayList");
        /*获取指定区间的元素,同时支持从后向前的下标,-1表示倒数第一个元素,-2表示倒数第二个元素,
以此类推*/
        System.out.println("myList的全部元素："
                          +jedis.lrange("myList", 0, -1));
        System.out.println("指定myList区间1~2的元素："
                          +jedis.lrange("myList",1,2));
        /*将列表指定的值删除 , 第二个参数用于指定参数的个数(List类型是可重复的列表),并且遵循"先进
后出"原则,后保存的数据优先被删除 */
        System.out.println("将myList的指定元素删除："
```

```java
                        +jedis.lrem("myList", 2, "HashMap"));
        System.out.println("myList的全部元素："
                        +jedis.lrange("myList", 0, -1));
        System.out.println("删除指定区间以外的元素："
                        +jedis.ltrim("myList", 0, 3));
        System.out.println("myList的全部元素："
                        +jedis.lrange("myList", 0, -1));
        //出栈操作，即取出最左端的元素，并将此元素移除
        System.out.println("myList列表出栈："+jedis.lpop("myList"));
        System.out.println("myList的全部元素："
                        +jedis.lrange("myList", 0, -1));
        //在List的最后添加元素，返回List所有元素的个数(添加操作完成之后)
        System.out.println("向myList添加元素："
                        +jedis.rpush("myList", "EnumMap"));
        System.out.println("myList的全部元素："
                        +jedis.lrange("myList", 0, -1));
        //取出列表最右端的元素，并将此元素移除
        System.out.println("myList列表出栈（右端）："+jedis.rpop("myList"));
        System.out.println("myList的全部元素："
                        +jedis.lrange("myList", 0, -1));
        System.out.println("修改myList指定下标1的内容："
                        +jedis.lset("myList", 1, "LinkedList"));
        System.out.println("myList的全部元素："
                        +jedis.lrange("myList", 0, -1));
        System.out.println("myList的元素个数："+jedis.llen("myList"));
        System.out.println("获取myList指定下标为2的元素内容："
                        +jedis.lindex("myList", 2));
        jedis.lpush("sorted", "3","6","2","0","7","4");
        System.out.println("sorted排序前："+jedis.lrange("sorted", 0, -1));
        /* 对列表元素进行排序，返回排序后的数据，但是排序并不能改变原列表的内容。
        如果是非number的数据，则会按照ASCII码进行排序 */
        System.out.println(jedis.sort("sorted"));
        System.out.println("sorted排序后："+jedis.lrange("sorted", 0, -1));
    }
```

List 操作测试结果如图 7-17 所示。

```
myList的全部元素：[ArrayList, Tree, Map, Set, List]
指定myList区间1~2的元素：[Tree, Map]
将myList的指定元素删除：0
myList的全部元素：[ArrayList, Tree, Map, Set, List]
删除指定区间以外的元素：OK
myList的全部元素：[ArrayList, Tree, Map, Set]
myList列表出栈：ArrayList
myList的全部元素：[Tree, Map, Set]
向myList添加元素：4
myList的全部元素：[Tree, Map, Set, EnumMap]
myList列表出栈（右端）：EnumMap
myList的全部元素：[Tree, Map, Set]
修改myList指定下标1的内容：OK
myList的全部元素：[Tree, LinkedList, Set]
myList的元素个数：3
获取myList指定下标为2的元素内容：Set
sorted排序前：[4, 7, 0, 2, 6, 3]
[0, 2, 3, 4, 6, 7]
sorted排序后：[4, 7, 0, 2, 6, 3]
```

图 7-17　List 操作测试结果

（3）Set 操作，代码如下。

```java
    @Test
```

```java
public void test5() {
    Jedis jedis = getJedis();
    jedis.flushDB();
    //返回添加成功后元素的个数
    System.out.println("向集合中添加一个或多个元素~~~~~~~~~~~~~~~~~~~~~");
    System.out.println(jedis.sadd("mySet",
                    "s1","s2","s4","s3","s0","s8","s7", "s5"));
    //Set中的元素不可重复，相同元素无法添加成功
    System.out.println(jedis.sadd("mySet", "s6"));
    System.out.println(jedis.sadd("mySet", "s6"));
    System.out.println("mySet的所有元素："+jedis.smembers("mySet"));
    //删除成功返回1，失败返回0
    System.out.println("删除一个元素s0："+jedis.srem("mySet", "s0"));
    System.out.println("mySet的所有元素："+jedis.smembers("mySet"));
    //返回删除成功的元素个数
    System.out.println("删除两个元素s7和s6："
                    +jedis.srem("mySet", "s7","s6"));
    System.out.println("mySet的所有元素："+jedis.smembers("mySet"));
    //返回被移除的元素
    System.out.println("随机移除集合中的一个元素："
                    +jedis.spop("mySet"));
    System.out.println("mySet的所有元素："+jedis.smembers("mySet"));
    System.out.println("mySet中包含元素的个数："+jedis.scard("mySet"));
    //判断Set中是否存在某元素，返回boolean值
    System.out.println("判断mySet中是否存在元素s3："
                    +jedis.sismember("mySet", "s3"));
    System.out.println("判断mySet中是否存在元素s1："
                    +jedis.sismember("mySet", "s1"));
    System.out.println("判断mySet中是否存在元素s5："
                    +jedis.sismember("mySet", "s5"));
    //两个Set之间的元素移动
    System.out.println(jedis.sadd("mySet1",
                    "s1","s2","s4","s3","s0","s8","s7", "s5"));
    System.out.println(jedis.sadd("mySet2",
                    "s1","s2","s4","s3","s0","s8"));
    System.out.println("将mySet1中的元素e1删除并存入mySet3中："
                    +jedis.smove("mySet1", "mySet3", "s1"));
    System.out.println("将mySet1中的元素e2删除并存入mySet3中："
                    +jedis.smove("mySet1", "mySet3", "s2"));
    System.out.println("mySet1中的所有元素："
                    +jedis.smembers("mySet1"));
    System.out.println("mySet3中的所有元素："
                    +jedis.smembers("mySet3"));
}
```

Set 操作测试结果如图 7-18 所示。

（4）Hash 操作，代码如下。

```java
@Test
public void test6() {
    Jedis jedis = getJedis();
    jedis.flushDB();
    Map<String,String> map = new HashMap<String,String>();
    map.put("k1","v1");
    map.put("k2","v2");
    map.put("k3","v3");
    map.put("k4","v4");
```

```java
//设置一个散列，并存入多个键/值对数据
jedis.hmset("hash",map);
//设置一个键/值对
jedis.hset("hash", "k5", "v5");
//返回Map<String,String>
System.out.println("获取Hash的所有键/值对："+jedis.hgetAll("hash"));
//返回Set<String>，类似于Java中Map的keySet方法
System.out.println("获取Hash的所有键："+jedis.hkeys("hash"));
//返回List<String>，类似于Java中Map的values方法
System.out.println("获取Hash的所有值："+jedis.hvals("hash"));
//Map增量操作，返回增量操作之后的值
System.out.println("将k6保存的值加上一个整数，如果k6不存在则添加k6："
                  +jedis.hincrBy("hash", "k6", 6));
System.out.println("获取Hash的所有键/值对："+jedis.hgetAll("hash"));
System.out.println("将k6保存的值加上一个整数，如果k6不存在则添加k6："
                  +jedis.hincrBy("hash", "k6", 3));
System.out.println("获取Hash的所有键/值对："+jedis.hgetAll("hash"));
System.out.println("删除一个或者多个键/值对："
                  +jedis.hdel("hash", "k2"));
System.out.println("获取Hash的所有键/值对："+jedis.hgetAll("hash"));
System.out.println("获取Hash中键/值对的个数："+jedis.hlen("hash"));
//判断散列中是否存在某个key，返回boolean值
System.out.println("判断Hash中是否存在k2："
                  +jedis.hexists("hash","k2"));
System.out.println("判断Hash中是否存在k3："
                  +jedis.hexists("hash","k3"));
//获取散列中某个key的值，不存在时返回null
System.out.println("获取Hash中的值："+jedis.hmget("hash","k3"));
//获取散列中多个key的值，不存在时返回null
System.out.println("获取Hash中的值："
                  +jedis.hmget("hash","k3","k7"));
}
```

```
向集合中输入一个或多个元素~~~~~~~~~~~~~~~~~~~~~
8
1
0
mySet的所有元素：[s7, s2, s8, s5, s1, s3, s0, s6, s4]
删除一个元素s0：1
mySet的所有元素：[s8, s5, s1, s3, s7, s6, s4, s2]
删除两个元素s7和s6：2
mySet的所有元素：[s8, s5, s4, s2, s3, s1]
随机移除集合中的一个元素：s4
mySet的所有元素：[s8, s5, s2, s3, s1]
mySet中包含元素的个数：5
判断mySet中是否存在元素s3：true
判断mySet中是否存在元素s1：true
判断mySet中是否存在元素s5：true
8
6
将mySet1中的元素e1删除并存入mySet3中：1
将mySet1中的元素e2删除并存入mySet3中：1
mySet1中的所有元素：[s7, s4, s8, s5, s0, s3]
mySet3中的所有元素：[s2, s1]
```

图 7-18　Set 操作测试结果

Hash 操作测试结果如图 7-19 所示。

```
获取Hash的所有键/值对：{k3=v3, k4=v4, k5=v5, k1=v1, k2=v2}
获取Hash的所有键  ：[k3, k4, k5, k1, k2]
获取Hash的所有值：[v3, v4, v2, v1, v5]
将k6保存的值加上一个整数，如果k6不存在则添加k6：6
获取Hash的所有键/值对：{k3=v3, k4=v4, k5=v5, k6=6, k1=v1, k2=v2}
将k6保存的值加上一个整数，如果k6不存在则添加k6：9
获取Hash的所有键/值对：{k3=v3, k4=v4, k5=v5, k6=9, k1=v1, k2=v2}
删除一个或者多个键/值对：1
获取Hash的所有键/值对：{k3=v3, k4=v4, k5=v5, k6=9, k1=v1}
获取Hash中键/值对的个数：5
判断Hash中是否存在k2：false
判断Hash中是否存在k3：true
获取Hash中的值：[v3]
获取Hash中的值：[v3, null]
```

图 7-19 Hash 操作测试结果

（5）Sorted Set 操作，代码如下。

```java
@Test
public void test7() {
    Jedis jedis = getJedis();
    jedis.flushDB();
    Map<String,Double> map = new HashMap<>();
    map.put("k3",4.0);
    map.put("k4",5.0);
    //为Sorted Set添加一个分数为3的元素，返回添加的元素个数
    System.out.println(jedis.zadd("mySorted", 3,"k1"));
    //向有序集合中添加多个元素及对应的分数，返回添加的元素个数
    System.out.println(jedis.zadd("mySorted",map));
    //获取所有元素
    System.out.println("获取mySorted中的元素：\r\n "
                +jedis.zrange("mySorted", 0, -1));
    //获取所有元素及元素对应的分数
    System.out.println("获取mySorted中的元素以及分数：\r\n "
            +jedis.zrangeWithScores("mySorted", 0, -1));
    System.out.println("获取mySorted中的所有元素：\r\n "
                +jedis.zrangeByScore("mySorted", 0,100));
    //获取指定分数范围内的元素
    System.out.println("获取mySorted中的元素：\r\n "
                +jedis.zrangeByScoreWithScores("mySorted", 0,100));
    System.out.println("获取mySorted中k2的分值："
                +jedis.zscore("mySorted", "k2"));
    System.out.println("获取mySorted中k2的排名："
                +jedis.zrank("mySorted", "k2"));
    //删除成功返回1，失败返回0
    System.out.println("将mySorted中的元素k3删除："
                +jedis.zrem("mySorted", "k3"));
    System.out.println("获取mySorted中的元素：\r\n "
                +jedis.zrange("mySorted", 0, -1));
    System.out.println("获取mySorted中元素的个数："
                +jedis.zcard("mySorted"));
    System.out.println("mySorted中分值在1～4之间的元素的个数："
                +jedis.zcount("mySorted", 1, 4));
    //分值增量操作
    System.out.println("key2的分值加上5："
                +jedis.zincrby("mySorted", 5, "k2"));
    System.out.println("key3的分值加上4："
```

```
                                        +jedis.zincrby("mySorted", 4, "k3"));
    System.out.println("mySorted中的所有元素：\r\n "
                                        +jedis.zrange("mySorted", 0, -1));
}
```

Sorted Set 操作测试结果如图 7-20 所示。

```
1
2
获取mySorted中的元素：
    [k1, k3, k4]
获取mySorted中的元素以及分数：
    [[[107, 49],3.0], [[107, 51],4.0], [[107, 52],5.0]]
获取mySorted中的所有元素：
    [k1, k3, k4]
获取mySorted中的元素：
    [[[107, 49],3.0], [[107, 51],4.0], [[107, 52],5.0]]
获取mySorted中k2的分值：null
获取mySorted中k2的排名：null
将mySorted中的元素k3删除：1
获取mySorted中的元素：
    [k1, k4]
获取mySorted中元素的个数：2
mySorted中分值在1~4之间的元素的个数：1
key2的分值加上5：5.0
key3的分值加上4：4.0
mySorted中的所有元素：
    [k1, k3, k2, k4]
```

图 7-20　Sorted Set 操作测试结果

（6）键操作，代码如下。

```
@Test
public void test8() throws InterruptedException {
    Jedis jedis = getJedis();
    jedis.flushDB();
    System.out.println("判断某个键是否存在："+jedis.exists("name"));
    System.out.println("新增键/值对："+jedis.set("name", "zhangsan"));
    //判断键是否存在
    System.out.println(jedis.exists("name"));
    System.out.println("新增键/值对："+jedis.set("pwd", "pwd"));
    System.out.println("获取所有的key：");
    Set<String> keys = jedis.keys("*");
    //循环输出所有的key
    keys.forEach((key) -> System.out.println( "        " + key));
    System.out.println("删除键pwd:"+jedis.del("pwd"));
    System.out.println("判断键pwd是否存在："+jedis.exists("pwd"));
    //设置过期时间，单位为s，返回1表示设置成功
    System.out.println("设置键name的过期时间为5s:"
                +(jedis.expire("name", 5)==1?"成功":"失败"));
    TimeUnit.SECONDS.sleep(2);
    //返回剩余时间，单位为s
    System.out.println("name的剩余生存时间："+jedis.ttl("name"));
    //撤销过期时间
    System.out.println("撤销name的过期时间："+jedis.persist("name"));
    //在2.8及以上版本中，返回-1表示没有设置过期时间，返回-2表示键不存在
    //在2.6或者更早的版本中，没有设置过期时间或键不存在时都返回-1
    System.out.println("name的剩余生存时间："+jedis.ttl("name"));
    System.out.println("name所存储的值的类型："+jedis.type("name"));
}
```

键操作测试结果如图 7-21 所示。

```
判断某个键是否存在：false
新增键/值对：OK
true
新增键/值对：OK
获取所有的key：
        name
        pwd
删除键pwd：1
判断键pwd是否存在：false
设置键name的过期时间为5s：成功
name的剩余生存时间：3
撤销name的过期时间：1
name的剩余生存时间：-1
name所存储的值的类型：string
```

图 7-21　键操作测试结果

7.2.2　Spring 与 Jedis 的集成

V7-2　Redis 在 JavaWeb 中的应用

随着用户量的提升，并发量提高，服务器压力变大，响应时间也变长。当访问网页、使用 App 的时候，速度、安全是用户考虑的重要问题，一个时延长的程序注定会流失用户；当然，不注重用户数据安全也注定会流失用户。在互联网界，此类案例不在少数。

Redis 通常不会单独使用，可以使 Spring 与 Jedis 集成使用，以方便对 Redis 的操作。由于 Redis 的特性，在 Web 项目中，Redis 通常作为缓存使用。

在某 Web 项目中，在注册时可能需要选择城市数据，通常城市数据比较多，若用户发起注册操作时再到 MySQL 中查询城市数据，在用户并发量不高的情况下，服务器还能够应付，但在高峰情况下，就是对系统性能的极大考验。此时，缓存的重要性就显现出来。因为像城市这样的数据，变化不会太大，通常做查询操作，在查询的时候，直接从 Redis 中加载数据即可。Redis 具有优秀的读写能力，能以超低延时返回数据。

关于 Spring 与 Redis 的集成，在前文中已有介绍，具体的配置步骤这里不再赘述，第 8 章将详细为读者介绍缓存的使用。

第 8 章

缓存

学习目标

（1）了解缓存实现方案。
（2）掌握Ehcache缓存应用。
（3）掌握如何在SSM项目中整合Redis实现缓存。

本章导读

■ 在前面的章节中，所有的数据都存储在数据库中，对数据的操作也直接转换为对数据库的操作。在高并发的情况下，数据库的频繁操作将导致整个系统运行缓慢，甚至变为不可用状态。本章将通过在数据库中添加缓存来解决高并发情况下应用卡顿的问题。

8.1 缓存实现方案

可以把缓存理解为在服务器运行期间存在于内存中的数据。数据类型并不一定需要强制要求，但键/值类型是比较好的选择。

下面通过编码实现简单的缓存机制。

```java
package com.woniu.cache;

import java.util.Collections;
import java.util.HashMap;
import java.util.Map;

public class Cache {
    //缓存对象，多线程操作情况下保证数据的安全性
    private Map<String,Object> localCacheStore =
                Collections.synchronizedMap(new HashMap<>());
    private static Cache cache = new Cache();
    //私有化构造器
    private Cache(){
    }
    //对外提供唯一的获取Cache的入口
    public static Cache getInstance(){
        return cache;
    }
    //获取
    public Object getValueByKey(String key){
        return localCacheStore.get(key);
    }
    //存储
    public void putValue(String key ,Object value){
        localCacheStore.put(key, value);
    }
}
```

如以上代码所示，创建了一个用于进行缓存操作的 Cache 类。使用该类可以进行简单的缓存数据操作。但是缓存存在于内存中，该 Cache 类并没有对数据时效性设置操作，服务器运行时间长时，内存泄露的危险性就很高。

为了规范 Java 中的缓存操作，在 JavaEE 7 中更新了 JSR107 标准，对 Java 中的缓存做出了规范，进行了缓存空间大小、缓存失效策略、数据持久化、数据统计等一系列设置。现在已经有很多开源框架提供了对缓存的支持。

在当下企业中使用的缓存主要有 Ehcache、Redis、Mecached 等。本章主要介绍 Ehcache 与 Redis 作为缓存的使用。

8.2 Ehcache 实现

缓存数据一般是系统中共用的、常用的数据。下面以中国行政区划作为基础数据，结合 MyBatis 进行缓存操作。

1. 数据创建

中国的行政单位数据比较多，为了提高效率，通常将其存储为 3 张或 4 张表并进行关联，由于篇幅的关系，这里仅展示部分数据的插入。

（1）省级行政单位表，其部分内容如下。

```sql
##省级行政单位表
DROP TABLE IF EXISTS 'provinces';
CREATE TABLE 'provinces' (
  'id' int(11) NOT NULL AUTO_INCREMENT,
  'provinceid' varchar(20) NOT NULL,
  'province' varchar(50) NOT NULL,
  PRIMARY KEY ('id')
) ENGINE=MyISAM AUTO_INCREMENT=35 DEFAULT CHARSET=utf8 COMMENT='省份信息表';
-- ----------------------------
-- Records of provinces
-- ----------------------------
INSERT INTO 'provinces' VALUES ('1', '110000', '北京市');
INSERT INTO 'provinces' VALUES ('2', '120000', '天津市');
INSERT INTO 'provinces' VALUES ('3', '130000', '河北省');
INSERT INTO 'provinces' VALUES ('4', '140000', '山西省');
INSERT INTO 'provinces' VALUES ('5', '150000', '内蒙古自治区');
INSERT INTO 'provinces' VALUES ('6', '210000', '辽宁省');
INSERT INTO 'provinces' VALUES ('7', '220000', '吉林省');
INSERT INTO 'provinces' VALUES ('8', '230000', '黑龙江省');
INSERT INTO 'provinces' VALUES ('9', '310000', '上海市');
INSERT INTO 'provinces' VALUES ('10', '320000', '江苏省');
INSERT INTO 'provinces' VALUES ('11', '330000', '浙江省');
INSERT INTO 'provinces' VALUES ('12', '340000', '安徽省');
INSERT INTO 'provinces' VALUES ('13', '350000', '福建省');
INSERT INTO 'provinces' VALUES ('14', '360000', '江西省');
INSERT INTO 'provinces' VALUES ('15', '370000', '山东省');
INSERT INTO 'provinces' VALUES ('16', '410000', '河南省');
INSERT INTO 'provinces' VALUES ('17', '420000', '湖北省');
INSERT INTO 'provinces' VALUES ('18', '430000', '湖南省');
INSERT INTO 'provinces' VALUES ('19', '440000', '广东省');
INSERT INTO 'provinces' VALUES ('20', '450000', '广西壮族自治区');
INSERT INTO 'provinces' VALUES ('21', '460000', '海南省');
INSERT INTO 'provinces' VALUES ('22', '500000', '重庆市');
```

（2）城市表，其部分内容如下。

```sql
##城市表
DROP TABLE IF EXISTS 'cities';
CREATE TABLE 'cities' (
  'id' int(11) NOT NULL AUTO_INCREMENT,
  'cityid' varchar(20) NOT NULL,
  'city' varchar(50) NOT NULL,
  'provinceid' varchar(20) NOT NULL,
  PRIMARY KEY ('id')
) ENGINE=MyISAM AUTO_INCREMENT=354
DEFAULT CHARSET=utf8 COMMENT='行政区域地州市信息表';
-- ----------------------------
-- Records of cities
-- ----------------------------
INSERT INTO 'cities' VALUES ('1', '110100', '市辖区', '110000');
INSERT INTO 'cities' VALUES ('2', '120100', '市辖区', '120000');
INSERT INTO 'cities' VALUES ('3', '130100', '石家庄市', '130000');
INSERT INTO 'cities' VALUES ('4', '130200', '唐山市', '130000');
INSERT INTO 'cities' VALUES ('5', '130300', '秦皇岛市', '130000');
INSERT INTO 'cities' VALUES ('6', '130400', '邯郸市', '130000');
##more
```

（3）县级表，其部分内容如下。

```sql
##县级表
DROP TABLE IF EXISTS 'areas';
CREATE TABLE 'areas' (
  'id' int(11) NOT NULL AUTO_INCREMENT,
  'areaid' varchar(20) NOT NULL,
  'area' varchar(50) NOT NULL,
  'cityid' varchar(20) NOT NULL,
  PRIMARY KEY ('id')
) ENGINE=MyISAM AUTO_INCREMENT=3303 DEFAULT CHARSET=utf8 COMMENT='行政区域县区信息表';
-- ----------------------------
-- Records of areas
-- ----------------------------
INSERT INTO 'areas' VALUES ('1', '110101', '东城区', '110100');
INSERT INTO 'areas' VALUES ('2', '110102', '西城区', '110100');
INSERT INTO 'areas' VALUES ('3', '110105', '朝阳区', '110100');
INSERT INTO 'areas' VALUES ('4', '110106', '丰台区', '110100');
INSERT INTO 'areas' VALUES ('5', '110107', '石景山区', '110100');
INSERT INTO 'areas' VALUES ('6', '110108', '海淀区', '110100');
INSERT INTO 'areas' VALUES ('7', '110109', '门头沟区', '110100');
INSERT INTO 'areas' VALUES ('8', '110111', '房山区', '110100');
INSERT INTO 'areas' VALUES ('9', '110112', '通州区', '110100');
INSERT INTO 'areas' VALUES ('10', '110113', '顺义区', '110100');
INSERT INTO 'areas' VALUES ('11', '110114', '昌平区', '110100');
INSERT INTO 'areas' VALUES ('12', '110115', '大兴区', '110100');
INSERT INTO 'areas' VALUES ('13', '110116', '怀柔区', '110100');
INSERT INTO 'areas' VALUES ('14', '110117', '平谷区', '110100');
INSERT INTO 'areas' VALUES ('15', '110118', '密云区', '110100');
##more
```

2. SSM 搭建

SSM 搭建使用 Spring Boot 的方式，在 pom.xml 中添加依赖。

（1）添加 Maven 依赖，代码如下。

```xml
<!-- 继承父Boot -->
<parent>
    <groupId>org.springframework.boot</groupId>
    <artifactId>spring-boot-starter-parent</artifactId>
    <version>1.5.2.RELEASE</version>
</parent>
<dependencies>
    <dependency>
        <groupId>org.springframework.boot</groupId>
        <artifactId>spring-boot-starter-web</artifactId>
    </dependency>
    <dependency>
        <groupId>org.springframework.boot</groupId>
        <artifactId>spring-boot-starter-test</artifactId>
        <scope>test</scope>
    </dependency>
    <dependency>
        <groupId>org.springframework.boot</groupId>
        <artifactId>spring-boot-devtools</artifactId>
    </dependency>
    <!--Spring Boot整合MyBatis -->
    <dependency>
```

```xml
        <groupId>org.mybatis.spring.boot</groupId>
        <artifactId>mybatis-spring-boot-starter</artifactId>
        <version>1.1.1</version>
    </dependency>
    <!-- MySQL驱动 -->
    <dependency>
        <groupId>mysql</groupId>
        <artifactId>mysql-connector-java</artifactId>
    </dependency>
    <!-- 基本配置 -->
    <dependency>
        <groupId>junit</groupId>
        <artifactId>junit</artifactId>
        <scope>test</scope>
    </dependency>
    <dependency>
        <groupId>org.springframework.boot</groupId>
        <artifactId>spring-boot-starter-tomcat</artifactId>
        <scope>provided</scope>
    </dependency>
    <dependency>
        <groupId>org.apache.tomcat.embed</groupId>
        <artifactId>tomcat-embed-jasper</artifactId>
        <scope>provided</scope>
    </dependency>
    <dependency>
        <groupId>javax.servlet</groupId>
        <artifactId>jsp-api</artifactId>
        <version>2.0</version>
        <scope>provided</scope>
    </dependency>
</dependencies>
<build>
    <!-- 插件 -->
    <plugins>
        <plugin>
            <groupId>org.springframework.boot</groupId>
            <artifactId>spring-boot-maven-plugin</artifactId>
        </plugin>
    </plugins>
</build>
```

（2）配置数据库连接。创建文件application.yml，用于存储数据库连接参数，内容如下。

```
spring:
  datasource:
    driver-class-name: com.mysql.jdbc.Driver
    url: jdbc:mysql://localhost/cache
    username: root
    password: root
```

（3）创建Mapper。ProvinceMapper接口的代码如下。

```
package com.woniu.dao.mapper;

import java.util.List;

import org.apache.ibatis.annotations.Mapper;
import org.apache.ibatis.annotations.Select;
```

```
import com.woniu.dao.entity.Province;

@Mapper
public interface ProvinceMapper {
    //获取所有的省
    @Select(value="select * from provinces")
    List<Province> getAllProvince();
}
```

CityMapper 接口的代码如下。

```
package com.woniu.dao.mapper;

import java.util.List;

import org.apache.ibatis.annotations.Mapper;
import org.apache.ibatis.annotations.Select;

import com.woniu.dao.entity.City;

@Mapper
public interface CityMapper {
    //根据省ID获取城市
    @Select(value="select * from cities where provinceid=${_parameter}")
    List<City> getCitysByProvinceId(String provinceId);
}
```

AreaMapper 接口的内容如下。

```
package com.woniu.dao.mapper;

import java.util.List;

import org.apache.ibatis.annotations.Mapper;
import org.apache.ibatis.annotations.Select;

import com.woniu.dao.entity.Area;

@Mapper
public interface AreaMapper {
    //根据城市ID获取地区数据
    @Select("select * from areas where cityid = #{_parameter}")
    List<Area> getAreaByCityid(String cityid);
}
```

Mapper 接口中的方法比较简单，SQL 使用 MyBatis 提供的注解进行开发，非常方便。接口中使用到的 Area、City、Province 类是 POJO 映射类。

（4）Service 编写。业务层分为实现层与接口层。接口 AddressService 的代码如下。

```
package com.woniu.service;

import java.util.List;

import com.woniu.dao.entity.Area;
import com.woniu.dao.entity.City;
import com.woniu.dao.entity.Province;

public interface AddressService {
    //获取所有的省
```

```java
    List<Province> getAllProvince();
    //根据城市ID获取地区数据
    List<Area> getAreaByCityid(String cityid);
    //根据省ID获取城市数据
    List<City> getCitysByProvinceId(String provinceId);
}
```

实现层需要使用Spring将Mapper接口注入Service，并调用Mapper接口处理数据，代码如下。

```java
package com.woniu.service;

import java.util.List;

import org.springframework.beans.factory.annotation.Autowired;
import org.springframework.stereotype.Service;

import com.woniu.dao.entity.*;
import com.woniu.dao.mapper.*;

@Service
public class AddressServiceImpl implements AddressService{
    @Autowired
    private AreaMapper areaMapper;
    @Autowired
    private CityMapper cityMapper;
    @Autowired
    private ProvinceMapper provinceMapper;
    @Override
    public List<Province> getAllProvince() {
        return provinceMapper.getAllProvince();
    }

    @Override
    public List<Area> getAreaByCityid(String cityid) {
        return areaMapper.getAreaByCityid(cityid);
    }

    @Override
    public List<City> getCitysByProvinceId(String provinceId) {
        return cityMapper.getCitysByProvinceId(provinceId);
    }

}
```

（5）编写Controller层。AddressController类用于处理地区信息加载，代码如下。

```java
package com.woniu.controller;

import java.util.List;

import org.springframework.beans.factory.annotation.Autowired;
import org.springframework.web.bind.annotation.RequestMapping;
import org.springframework.web.bind.annotation.RestController;

import com.woniu.dao.entity.Area;
import com.woniu.dao.entity.City;
import com.woniu.dao.entity.Province;
import com.woniu.service.AddressService;
import com.woniu.vo.Result;
```

```java
@RestController
@RequestMapping("addr")
public class AddressController extends BaseController{
    @Autowired
    private AddressService addressService;
    //获取省数据
    @RequestMapping("getProvince")
    public Result getProvince() {
        List<Province> allProvince = addressService.getAllProvince();
        if(allProvince==null || allProvince.size()==0) {
            return none();
        }
        return ok(allProvince);
    }
    //获取城市数据
    @RequestMapping("getCities")
    public Result getCities(String provinceId) {
        List<City> citysByProvinceId =
                addressService.getCitysByProvinceId (provinceId);
        if(citysByProvinceId==null || citysByProvinceId.size()==0) {
            return none();
        }
        return ok(citysByProvinceId);
    }
    //获取地区
    @RequestMapping("getArea")
    public Result getArea(String cityid) {
        List<Area> areaByCityid = addressService.getAreaByCityid(cityid);
        if(areaByCityid==null || areaByCityid.size()==0) {
            return none();
        }
        return ok(areaByCityid);
    }
}
```

AddressController 类继承自 BaseController，其主要用于提供基础 Controller 类需要的一些通用方法。

```java
package com.woniu.controller;

import com.woniu.vo.Result;
//基础Controller功能
public class BaseController {
    //成功返回
    public Result ok(Object obj) {
        return new Result("0", "", obj);
    }
    //成功
    public Result ok() {
        return ok(null);
    }
    //错误
    public Result error() {
        return new Result("error", "发生错误", null);
    }
    //未找到
    public Result none() {
```

```
        return new Result("none", "找不到相应数据", null);
    }
}
```

在实际项目开发中，通常会使用AJAX（Asynchronous JavaScript And XML，异步JavaScript和XML）传输JSON数据并进行交互，且以通用的JSON数据格式作为返回值。Result类是本项目返回信息的实体类，只需要在处理请求的方法上添加ResponseBody注解，Spring就会自动将数据转换为JSON格式。Spring也提供了类级别的ResponseBody注解，即RestController。该注解有两个作用：一是包含Controller注解，如果添加了该注解，则不需要再添加Controller注解；二是表示当前类中所有的处理请求方法都返回JSON数据，也不需要再添加ResponseBody注解。Result的代码如下。

```
//统一JSON格式
public class Result {
    //处理请求的状态码，0表示成功，1表示失败
    private String code;
    private String msg;
    //返回的数据
    private Object data;
    //getter/setter方法
```

在系统中，不可能所有的请求都返回JSON数据，也会进行页面转发、重定向等，可以专门写一个路径跳转类UrlController，代码如下。

```
package com.woniu.controller;

import org.springframework.stereotype.Controller;
import org.springframework.web.bind.annotation.RequestMapping;
@Controller
public class UrlController {
    //页面跳转
    @RequestMapping("index")
    public String index() {
        return "/addr";
    }
}
```

（6）Spring Boot配置。使用Spring提供的Config形式进行配置，代码如下。

```
package com.woniu;

import org.springframework.context.annotation.Bean;
import org.springframework.context.annotation.Configuration;
import org.springframework.web.servlet.ViewResolver;
import org.springframework.web.servlet.config.annotation.EnableWebMvc;
import org.springframework.web.servlet.config.annotation.WebMvcConfigurerAdapter;
import org.springframework.web.servlet.view.InternalResourceViewResolver;
//基础数据配置类
@Configuration
@EnableWebMvc
public class ViewConfig extends WebMvcConfigurerAdapter{
    @Bean
    public ViewResolver getView() {
        InternalResourceViewResolver resolver = new InternalResourceViewResolver();
        //设置视图前、后缀
        resolver.setPrefix("/page");
        resolver.setSuffix(".jsp");
        return resolver;
    }
}
```

Spring Boot 启动的入口 Application 的内容如下。

```java
package com.woniu;

import org.springframework.boot.SpringApplication;
import org.springframework.boot.autoconfigure.SpringBootApplication;
//启动入口
@SpringBootApplication
public class Application {
    public static void main(String[] args) {
        SpringApplication.run(Application.class, args);
    }
}
```

（7）测试页面 addr.jsp，代码如下。

```jsp
<%@ page language="java" contentType="text/html; charset=utf-8"
    pageEncoding="utf-8"%>
<%
String path = request.getContextPath();
String basePath = request.getScheme()+
"://"+request.getServerName()+":"+request.getServerPort()+path+"/";
%>
<!DOCTYPE html>
<html>
<head>
<meta http-equiv="Content-Type" content="text/html; charset=utf-8">
<script src="http://code.jquery.com/jquery-2.1.4.min.js"></script>
<title>Insert title here</title>
</head>
<body>
    请选择地区：<select id="province" onchange="loadCity()">
            <option value="" selected="selected">请选择省</option>
        </select>
        <select id="city" onchange="loadArea()">
            <option value="">请选择城市</option>
        </select>
        <select id="area">
            <option value="">请选择区域</option>
        </select>
</body>
<script type="text/javascript">
    $(function(){
        loadProv();
    })
    /*加载省数据*/
    function loadProv(){
        $.ajax({
            url:"/addr/getProvince",
            type:"post",
            dataType:"json",
            success:function(data){
                if(data.code=='0'){
                    var proStr = '';
                    $.each(data.data,function(i,item){
                        proStr += '<option value="'+item.provinceid+'">'+item.province+'</option>';
                    });
```

```javascript
                $("#province").append(proStr);
                return;
            }
            alert(data.msg);
        }
    })
}
/*根据省ID加载城市数据*/
function loadCity(){
    var province = $("#province").val();
    $.ajax({
        url:"/addr/getCities",
        data:{"provinceId":province},
        type:"post",
        dataType:"json",
        success:function(data){
            if(data.code=='0'){
                $("#city").empty();
                var cityStr = '';
                $.each(data.data,function(i,item){
                    cityStr += '<option
 value="'+item.cityid+'">'+item.city+'</option>';
                });
                $("#city").append(cityStr);
                return;
            }
            alert(data.msg);
        }
    })
}
/*根据城市ID加载地区数据*/
function loadArea(){
    var cityid = $("#city").val();
    $.ajax({
        url:"/addr/getArea",
        type:"post",
        data:{"cityid":cityid},
        dataType:"json",
        success:function(data){
            if(data.code=='0'){
                $("#area").empty();
                var areaStr = '';
                $.each(data.data,function(i,item){
                    areaStr += '<option
 value="'+item.areaid+'">'+item.area+'</option>';
                });
                $("#area").append(areaStr);
                return;
            }
            alert(data.msg);
        }
    })
}
</script>
</html>
```

运行 Application 类，访问 http://localhost:8080/index，测试结果如图 8-1 所示。

图 8-1　测试结果

为了更好地检测测试结果，配置 Spring Boot 日志级别为 DEBUG，输出 SQL 语句。创建文件 application.properties，添加如下内容。

```
### Debug
logging.level.com.woniu=DEBUG
```

当选择省市区的时候，可以看到控制台会一直输出 SQL 语句，如图 8-2 所示。

```
==> Parameters: 150300(String)
<==      Total: 4
==>  Preparing: select * from areas where cityid = ?
==> Parameters: 150400(String)
<==      Total: 13
==>  Preparing: select * from areas where cityid = ?
==> Parameters: 150900(String)
<==      Total: 12
==>  Preparing: select * from cities where provinceid=210000
==> Parameters:
<==      Total: 14
==>  Preparing: select * from areas where cityid = ?
==> Parameters: 210800(String)
<==      Total: 7
==>  Preparing: select * from areas where cityid = ?
==> Parameters: 211000(String)
<==      Total: 8
==>  Preparing: select * from areas where cityid = ?
==> Parameters: 211200(String)
<==      Total: 8
==>  Preparing: select * from cities where provinceid=410000
==> Parameters:
<==      Total: 18
==>  Preparing: select * from cities where provinceid=430000
==> Parameters:
<==      Total: 14
```

图 8-2　控制台输出 SQL 语句

控制台输出 SQL 语句，说明在用户选择时，会到数据库中进行访问，并返回数据，这样的操作会使数据库的压力非常大。可以通过在系统中加入缓存的方法来解决这个问题。

3. Ehcache 配置

（1）引入 Ehcache。在 pom.xml 文件中添加如下依赖。

```xml
<dependency>
    <groupId>org.springframework.boot</groupId>
    <artifactId>spring-boot-starter-cache</artifactId>
</dependency>
<dependency>
    <groupId>net.sf.ehcache</groupId>
    <artifactId>ehcache</artifactId>
</dependency>
```

（2）配置 Ehcache 文件。创建 ehcache.xml 文件，将其放置于根目录中，并添加相关配置，代码如下。

```xml
<?xml version="1.0" encoding="UTF-8"?>
<ehcache xmlns:xsi="http://www.w3.org/2001/XMLSchema-instance"
    xsi:noNamespaceSchemaLocation="http://ehcache.org/ehcache.xsd"
    updateCheck="false">
    <diskStore path="user.dir/sqlEhCache" />
    <!-- 默认缓存 -->
    <defaultCache eternal="false" maxElementsInMemory="1000"
        overflowToDisk="false" diskPersistent="false" timeToIdleSeconds="0"
```

```
                timeToLiveSeconds="600" memoryStoreEvictionPolicy="LRU" />
        <!-- 配置缓存,需要注意的是name属性,在使用缓存的时候需要该属性 -->
        <cache name="addrCache" eternal="true"
            maxElementsInMemory="1000"
            maxElementsOnDisk="10000"
            overflowToDisk="true"
            diskPersistent="false" timeToIdleSeconds="0"
            timeToLiveSeconds="300"
            memoryStoreEvictionPolicy="LRU" />
</ehcache>
```

在 application.properties 文件中配置缓存文件的位置。

```
spring.cache.ehcache.config=classpath:ehcache.xml
```

在 AddressServiceIpl 类中需要缓存的方法上添加缓存注解。

```
@Cacheable(value="addrCache")
```

value 属性就是在 ehcache.xml 文件中配置的 name 属性。在 Application 类上添加 EnableCaching 开启允许缓存功能。

Cacheable 注解的作用是在进行查询时,优先到缓存中查找,没有查找到数据则到数据库中查找。正如测试结果所示,当第一次获取某地区数据时,Eclipse 控制台会输出 SQL 语句,而再次单击获取该地区的数据时,没有输出 SQL 语句,数据库的压力大大减小了,缓存测试结果如图 8-3 所示。

```
~~~~~~获取省110000的城市数据
==>  Preparing: select * from cities where provinceid=110000
==> Parameters:
<==      Total: 1
~~~~~~获取省120000的城市数据
==>  Preparing: select * from cities where provinceid=120000
==> Parameters:
<==      Total: 1
~~~~~~获取省130000的城市数据
==>  Preparing: select * from cities where provinceid=130000
==> Parameters:
<==      Total: 12
~~~~~~获取省110000的城市数据
~~~~~~获取省120000的城市数据
~~~~~~获取省130000的城市数据
```

图 8-3 缓存测试结果

8.3 Redis 实现

V8-1 缓存的基本操作

Redis 与 Spring Boot 及 MyBatis 进行集成的功能和 Ehcache 基本一致,都是作为 MyBatis 的二级缓存使用,以缓解数据库的 I/O 瓶颈。

1. 创建 Maven 项目并加入依赖

创建 Maven 项目 spring-redis,加入 Spring Boot、MyBatis、Jedis 依赖,代码如下。

```
    <parent>
        <groupId>org.springframework.boot</groupId>
        <artifactId>spring-boot-starter-parent</artifactId>
        <version>1.5.2.RELEASE</version>
    </parent>
    <dependencies>
      <dependency>
          <groupId>org.springframework.boot</groupId>
          <artifactId>spring-boot-starter-web</artifactId>
      </dependency>

      <dependency>
```

```xml
        <groupId>org.springframework.boot</groupId>
        <artifactId>spring-boot-starter-redis</artifactId>
        <version>1.4.3.RELEASE</version>
</dependency>

<dependency>
    <groupId>com.alibaba</groupId>
    <artifactId>druid</artifactId>
    <version>1.0.29</version>
</dependency>
  <dependency>
        <groupId>org.springframework.boot</groupId>
        <artifactId>spring-boot-starter-test</artifactId>
        <scope>test</scope>
</dependency>
  <dependency>
        <groupId>org.springframework.boot</groupId>
        <artifactId>spring-boot-devtools</artifactId>
</dependency>
<!--Spring Boot整合MyBatis -->
<dependency>
        <groupId>org.mybatis.spring.boot</groupId>
        <artifactId>mybatis-spring-boot-starter</artifactId>
        <version>1.1.1</version>
</dependency>
<!-- MySQL驱动 -->
<dependency>
    <groupId>mysql</groupId>
    <artifactId>mysql-connector-java</artifactId>
  </dependency>
    <!-- 基本配置 -->
<dependency>
    <groupId>junit</groupId>
    <artifactId>junit</artifactId>
    <scope>test</scope>
</dependency>
<dependency>
        <groupId>org.springframework.boot</groupId>
        <artifactId>spring-boot-starter-tomcat</artifactId>
        <scope>provided</scope>
</dependency>
<dependency>
        <groupId>org.apache.tomcat.embed</groupId>
        <artifactId>tomcat-embed-jasper</artifactId>
        <scope>provided</scope>
</dependency>
  <dependency>
     <groupId>javax.servlet</groupId>
     <artifactId>jsp-api</artifactId>
     <version>2.0</version>
     <scope>provided</scope>
  </dependency>
  <dependency>
     <groupId>org.springframework.boot</groupId>
     <artifactId>spring-boot-starter-logging</artifactId>
  </dependency>
  </dependencies>
<build>
```

```xml
            <!-- 插件 -->
            <plugins>
                <plugin>
                    <groupId>org.springframework.boot</groupId>
                    <artifactId>spring-boot-maven-plugin</artifactId>
                </plugin>
            </plugins>
    </build>
```

2. Redis 连接

创建 application.yml 文件，对 Redis 连接进行配置，代码如下。

```yaml
spring:
  datasource:
    driver-class-name: com.mysql.jdbc.Driver
    url: jdbc:mysql://localhost/cache
    username: root
    password: root
  redis:
      database: 0
      host: 127.0.0.1
      port: 6379
      password: 123456
      max-active: 8
      max-wait: -1
      max-idle: 8
      min-idle: 0
      timeout: 0
```

3. Redis 配置

创建 RedisCacheConfig 类，对 Redis 进行基础配置，代码如下。与 Ehcache 不同的是，Ehcache 可以对没有参数的方法进行自动缓存，而 Redis 对没有参数的方法需要程序员提供主键生成器，否则会抛出异常。

```java
package com.woniu.config;

import org.springframework.cache.CacheManager;
import org.springframework.cache.annotation.CachingConfigurerSupport;
import org.springframework.cache.annotation.EnableCaching;
import org.springframework.cache.interceptor.KeyGenerator;
import org.springframework.context.annotation.Bean;
import org.springframework.context.annotation.Configuration;
import org.springframework.data.redis.cache.RedisCacheManager;
import org.springframework.data.redis.connection.RedisConnectionFactory;
import org.springframework.data.redis.core.RedisTemplate;
import org.springframework.data.redis.core.StringRedisTemplate;

//Redis配置类
@Configuration
@EnableCaching   //开启注解
public class RedisCacheConfig extends CachingConfigurerSupport {
    //CacheManager对象
    @Bean
    public CacheManager cacheManager(RedisTemplate<?, ?> redisTemplate) {
        CacheManager cacheManager = new RedisCacheManager(redisTemplate);
        return cacheManager;
    }
    /*RedisTemplate对象，在开发中可以使用硬编码的形式完成缓存的操作，此时，
    需要使用RedisTemplate对象*/
```

```java
@Bean
public StringRedisTemplate stringRedisTemplate(RedisConnectionFactory factory) {
    StringRedisTemplate stringRedisTemplate = new StringRedisTemplate();
    stringRedisTemplate.setConnectionFactory(factory);
    return stringRedisTemplate;
}
//主键生成器，用于对没有参数的方法进行缓存，使用注解Redis会自动判断方法是否有参数
@Bean
@Override
public KeyGenerator keyGenerator() {
    return (target, method, params) -> {
        StringBuilder sb = new StringBuilder();
        sb.append(target.getClass().getName());
        sb.append(method.getName());
        for (Object obj : params) {
            sb.append(obj.toString());
        }
        return sb.toString();
    };
}
```

4．测试

其他内容从 8.2 节 Ehcache 的演示项目中复制即可，这里不再赘述。

8.4 其他缓存操作

前面分别演示了 Ehcache 和 Redis 在项目中作为缓存时的使用方法，但是仅演示了查询功能的缓存。在实际的项目开发中，增、删、查、改是最基本的功能。例如，更新某条数据后，如果没有更新缓存，那么缓存中的数据与数据库中的数据会不一样，这会造成数据的紊乱。又例如，删除一条数据后，如果没有将缓存中的数据清除，在查询时，遵循优先取缓存的原则，会取出已经被删除的数据，即用户删除了数据，但仍可以查询出此数据，在系统中是无法接受这样的逻辑错误的。

Spring 提供了完整的注解实现缓存机制，这里介绍几个常用的注解。

（1）Cacheable 注解

此注解是在 Ehcache 及 Redis 演示中用到的注解，表示缓存优先，通常用于查询，查询时会查看缓存中是否有匹配的数据，有则直接取出数据，没有则从数据库中取出数据，并在缓存中存储一份数据。

（2）CachePut 注解

此注解不会从缓存中取出数据，而是从数据库中取出数据并更新到缓存中，常用于更新操作。当用户进行更新操作的时候，其可以保证缓存与数据库中的数据是一致的。

（3）CacheEvict 注解

此注解用于清除缓存，在删除数据的时候使用，用户进行删除操作后，此注解会将缓存中的数据删除，当再次查询此数据的时候，缓存中将查询不到此数据。

下面将商品的管理功能结合 Redis 缓存进行操作。Redis 基本配置沿用 8.3 节所用配置。

1．创建商品表

具体代码如下。

```sql
-- ----------------------------
-- 创建商品表gods
-- ----------------------------
DROP TABLE IF EXISTS 'gods';
CREATE TABLE 'gods' (
  'id' int(11) NOT NULL AUTO_INCREMENT,
```

```sql
  'gods_name' varchar(255) DEFAULT NULL COMMENT '商品名称',
  'gods_price' decimal(10,0) DEFAULT NULL COMMENT '商品价格',
  'gods_detail' varchar(255) DEFAULT NULL COMMENT '商品详情',
  PRIMARY KEY ('id')
) ENGINE=InnoDB AUTO_INCREMENT=21 DEFAULT CHARSET=utf8;

-- ----------------------------
-- 插入数据
-- ----------------------------
INSERT INTO 'gods' VALUES ('1', '智能语音助手', '999', '智能语音助手，陪您聊天');
INSERT INTO 'gods' VALUES ('2', '电饭煲', '499', '智能控火电饭煲');
INSERT INTO 'gods' VALUES ('3', '拖地王', '1999', '智能扫地机器人，清洁无烦恼');
INSERT INTO 'gods' VALUES ('4', '沙发', '2999', '高级沙发');
INSERT INTO 'gods' VALUES ('5', '大床', '2999', '高级大床');
INSERT INTO 'gods' VALUES ('6', '橱柜', '1999', '高级橱柜');
INSERT INTO 'gods' VALUES ('7', '冰箱', '2999', '智能冰箱');
INSERT INTO 'gods' VALUES ('8', '空调', '2999', '节能空调');
INSERT INTO 'gods' VALUES ('9', '吊灯', '1999', '客厅吊灯');
INSERT INTO 'gods' VALUES ('10', '电视机', '2999', '大屏电视机');
INSERT INTO 'gods' VALUES ('11', '笔记本电脑', '4999', '高级笔记本电脑');
INSERT INTO 'gods' VALUES ('12', '手机', '3999', '智能手机');
INSERT INTO 'gods' VALUES ('13', '打印机', '1999', '办公打印机');
INSERT INTO 'gods' VALUES ('14', '衣柜', '1999', '高级大衣柜');
INSERT INTO 'gods' VALUES ('15', '风扇', '399', '智能风扇');
INSERT INTO 'gods' VALUES ('16', '茶几', '499', '茶几');
INSERT INTO 'gods' VALUES ('17', '4件套', '599', '床上用品4件套，大甩卖');
INSERT INTO 'gods' VALUES ('18', '床垫', '599', '软床垫');
INSERT INTO 'gods' VALUES ('19', '圆桌', '899', '客厅圆桌');
INSERT INTO 'gods' VALUES ('20', '板凳', '99', '板凳');
```

2．编写 Mapper 接口

具体代码如下。

```java
package com.woniu.dao.mapper;

import java.util.List;

import org.apache.ibatis.annotations.*;

import com.woniu.dao.entity.Gods;

@Mapper
public interface GodsMapper {
    //分页查询
    @Select(value="select * from gods")
    public List<Gods> list();
    //更新商品
    public int updateGods(Gods gods);
    //删除商品
    @Delete(value="delete from gods where id=#{id}")
    public int delGods(Integer id);
    //根据主键查询商品
    @Select(value="select * from gods where id = #{id}")
    public Gods get(Integer id);
}
```

简单的 SQL 语句可以直接使用注解的形式完成；对于复杂的 SQL 语句，MyBatis 虽然也提供了类似的方式，但是通常不会使用，因为其书写上太乱。一般来说，在实际开发中，编写 Mapper 接口时，通常会使用注解与 Mapper 文件结合的形式。

GodsMapper.xml 文件的内容如下。

```xml
<?xml version="1.0" encoding="UTF-8" ?>
<!DOCTYPE mapper PUBLIC "-//mybatis.org//DTD Mapper 3.0//EN"
"http://mybatis.org/dtd/ mybatis-3-mapper.dtd" >
<mapper namespace="com.woniu.dao.mapper.GodsMapper" >
    <update id="updateGods" parameterType="com.woniu.dao.entity.Gods">
        update gods
            <set>
                <if test="godsName!=null">
                    gods_name = #{godsName},
                </if>
                <if test="godsPrice!=null">
                    gods_price = #{godsPrice}
                </if>
                <if test="godsPrice!=null">
                    gods_detail = #{godsDetail}
                </if>
            </set>

    </update>
</mapper>
```

3. Service 接口与实现

接口 GodsService 的内容如下。

```java
package com.woniu.service;

import com.github.pagehelper.PageInfo;
import com.woniu.dao.entity.Gods;

public interface GodsService {
    //查询商品列表(分页)
    public PageInfo<?> list(Integer pageNum, Integer pageSize);
    //更新商品信息
    public int updateGods(Gods gods);
    //删除商品信息
    public int deleteGods(Integer id);
    //获取商品信息
    public Gods get(Integer id);
}
```

GodsServiceImpl 实现类的内容如下。

```java
package com.woniu.service;


import org.springframework.beans.factory.annotation.Autowired;
import org.springframework.stereotype.Service;

import com.github.pagehelper.PageHelper;
import com.github.pagehelper.PageInfo;
import com.woniu.dao.entity.Gods;
import com.woniu.dao.mapper.GodsMapper;
@Service
public class GodsServiceImpl implements GodsService{
    @Autowired
    private GodsMapper godsMapper;
```

```java
    @Override
    public PageInfo<Gods> list(Integer pageNum, Integer pageSize) {
        //设置分页,pageNum表示当前页,pageSize表示每页的记录数量
        PageHelper.startPage(pageNum, pageSize);
        PageInfo<Gods> info = new PageInfo<>(godsMapper.list());
        return info;
    }
    @Override
    public int updateGods(Gods gods) {
        return godsMapper.updateGods(gods);
    }
    @Override
    public int deleteGods(Integer id) {
        return godsMapper.delGods(id);
    }
    @Override
    public Gods get(Integer id) {
        return godsMapper.get(id);
    }
}
```

查询商品的信息时需要进行分页。分页查询的逻辑非常简单,可以使用自定义的分页进行查询,也可以使用一些简单的插件进行查询。PageHelper 是为 MyBatis 提供分页支持的优秀插件,其还为 Spring Boot 提供了 starter。使用 Spring Boot 集成 PageHelper 框架的步骤如下。

(1)导入 Maven 依赖,代码如下。

```xml
<dependency>
    <groupId>com.github.pagehelper</groupId>
    <artifactId>pagehelper-spring-boot-starter</artifactId>
    <version>1.2.3</version>
</dependency>
```

(2)分页。PageHelper 框架有两个 API 类:一个是 PageHelper,用于设置分页的信息;另一个是 PageInfo,即分页的信息及数据,包括设置当前页、下一页等。分页的功能不算复杂,通过本节案例即可掌握 PageHelper 框架的基本使用方法。

4. 编写 Controller 类

GodsController 类的代码如下。

```java
package com.woniu.controller;

import org.springframework.beans.factory.annotation.Autowired;
import org.springframework.stereotype.Controller;
import org.springframework.ui.ModelMap;
import org.springframework.web.bind.annotation.PathVariable;
import org.springframework.web.bind.annotation.RequestMapping;
import org.springframework.web.bind.annotation.ResponseBody;

import com.github.pagehelper.PageInfo;
import com.woniu.dao.entity.Gods;
import com.woniu.service.GodsService;
import com.woniu.vo.Result;

@Controller
public class GodsController extends BaseController{
    @Autowired
    private GodsService godsService;
    //分页列表查询
```

```java
    @ResponseBody
    @RequestMapping("list")
    public PageInfo<?> list(Integer pageNum, Integer pageSize, ModelMap model) {
        PageInfo<?> list = godsService.list(pageNum, pageSize);
        return list;
    }
    //入口
    @RequestMapping("listPage")
    public String listPage() {
        return "/gods";
    }
    //跳转到修改页面
    @RequestMapping("toChange/{id}")
    public String toChange(@PathVariable Integer id, ModelMap model) {
        Gods gods = godsService.get(id);
        model.put("gods", gods);
        return "/updateGods";
    }
    //修改数据
    @ResponseBody
    @RequestMapping("change")
    public Result change(Gods gods) {
        System.out.println(gods);
        int updateGods = godsService.updateGods(gods);
        if(updateGods==0) {
            return error();
        }
        return ok();
    }
    //删除数据
    @ResponseBody
    @RequestMapping("del/{id}")
    public Result del(@PathVariable Integer id) {
        int deleteGods = godsService.deleteGods(id);
        if(deleteGods==0) {
            return error();
        }
        return ok();
    }
}
```

5. 编写 JSP 页面

在 JSP 页面中会使用分页功能，本案例使用 Bootstrap 进行编写。Gods.jsp 的代码如下。

```
<!DOCTYPE html>
<html>
<head>
<meta charset="utf-8">
<!-- 引入CSS样式 -->
<link rel="stylesheet" href="/static/css/qunit-1.11.0.css">
<link rel="stylesheet" href="/static/css/bootstrapv3.min.css">
<!-- JS -->
<script src="/static/js/jquery.min.js" type="text/javascript"></script>
<script src="/static/js/bootstrapv3.js" type="text/javascript"></script>
<script src="/static/js/bootstrap-paginator.js"></script>
<script src="/static/js/qunit-1.11.0.js"></script>
</head>
```

```html
<body>
    <div class="panel-body">
        <div class="table-responsive">
            <table class="table table-bordered">
                <thead>
                    <tr>
                        <th>序号</th>
                        <th>商品名称</th>
                        <th>价格</th>
                        <th>描述</th>
                        <th>操作</th>
                    </tr>
                </thead>
                <tbody id="tbody">

                </tbody>
            </table>
        </div>
        <div style="text-align: center;">
            <ul id='bp-3-element-test'></ul>
        </div>
    </div>
</body>
<script type="text/javascript">
    //初始化数据，在进行分页查询时，数据会发生改变
    localPageSize = 10;   //每页显示的数据条数
    var currentPage = 1;
    var numberOfPages = 5;
    var totalPages  = 1;
    $(function() {
        load_gods(1,localPageSize);
    });
    //加载商品(分页)
    load_gods = function(pageNum,pageSize){
        $.ajax({
            url:"list",
            type:"post",
            data:{
                "pageNum":pageNum,
                "pageSize":pageSize
            },
            dataType:"json",
            success:function(data){
                console.log(data);
                totalPages =  data.pages;
                currentPage = pageNum;
                if(data.pages<5){
                    numberOfPages = 5;
                } else {
                    numberOfPages = data.pages;
                }
                var str = "";
                $("#tbody").empty();
                $.each(data.list,function(index,item){
                    str += '<tr id="id'+item.id+'">'
                        +'<th>'+(index+1)+'</th>'
                        +'<th>'+item.godsName+'</th>'
```

```javascript
                            +'<th>'+item.godsPrice+'</th>'
                            +'<th>'+item.godsDetail+'</th>'
                            +'<th><button class="btn btn-warning" '
                            +'onclick="del('+item.id+')">删除</button>'
                            +'<button class="btn btn-info"'
                            +' onclick="change('+item.id+')">修改</button></th>'
                            +'</tr>';
                    });
                    $("#tbody").append(str);
                    initPager();
            }
        });
    }
    var initPager = function(){
        test("gods", function() {
            var element = $('#bp-3-element-test');
            var options = {
                bootstrapMajorVersion : 3,    //设置版本
                currentPage : currentPage,    //当前页数
                numberOfPages : numberOfPages,    //导航条显示的页数
                totalPages : totalPages,        //总页数
                itemTexts : function(type, page, current) {    //自定义导航条显示内容
                    switch (type) {
                    case "first":
                        return "首页";
                    case "prev":
                        return "上一页";
                    case "next":
                        return "下一页";
                    case "last":
                        return "末页";
                    case "page":
                        return page;
                    }
                },
                //导航条单击事件,page是分页需要的参数
                onPageClicked : function(event, originalEvent, type, page) {
                    load_gods(page,localPageSize);
                }
            }
            element.bootstrapPaginator(options);
        })
    };
    //修改
    var change = function(id){
        window.location = "/toChange/"+id;
    }
    //删除
    var del = function(id){
        $.ajax({
            url:"del/" + id,
            type:"post",
            dataType:"json",
            success:function(data){
                if(data.code=="0"){
                    alert("删除成功");
```

```
                    window.location = "/listPage";
                }
            }
        })
    }
    </script>
</body>
</html>
```

updateGods.jsp 是更新商品页面，其代码如下。

```jsp
<%@ page language="java" contentType="text/html; charset=utf-8"
    pageEncoding="utf-8"%>
<!DOCTYPE html>
<html>
<head>
<meta charset="utf-8">
<title>修改商品</title>
</head>
<body>
    <input type="hidden" name="id" value="${gods.id }" id="id">
    <div class="form-group">
      <label for="name">商品名称</label>
      <input type="text" class="form-control" id="godsName" placeholder="请输入商品名称" value="${gods.godsName }">
    </div>
     <div class="form-group">
      <label for="name">价格</label>
      <input type="text" class="form-control" id="godsPrice" placeholder="请输入商品价格" value="${gods.godsPrice }">
    </div>
     <div class="form-group">
      <label for="name">描述</label>
      <input type="text" class="form-control" id="godsDetail" placeholder="请输入商品描述" value="${gods.godsDetail }">
    </div>
    <button type="button" class="btn btn-default" onclick="return submit()"> 提 交 </button>
</body>
<script src="/static/js/jquery.min.js" type="text/javascript"></script>
<script type="text/javascript">
    //表单提交
    function submit(){
        var godsName = $("#godsName").val();
        var godsPrice = $("#godsPrice").val();
        var godsDetail = $("#godsDetail").val();
        if(godsName==null || godsName==""){
            alert("商品名称不能为空");
            return false;
        }
        if(godsPrice==null || godsPrice==""){
            alert("商品价格不能为空");
            return false;
        }
        if(godsDetail==null || godsDetail==""){
            alert("商品描述不能为空");
            return false;
        }
```

```
        console.log(godsDetail);
        $.ajax({
            url:"/change",
            //contentType: "application/json; charset=utf-8",
            data:{
                "godsName":godsName,
                "godsPrice":godsPrice,
                "godsDetail":godsDetail,
                "id":$("#id").val()
            },
            dataType:"json",
            type:"post",
            success:function(data){
                if(data.code=='0'){
                    window.location = '/listPage';
                }
                return false;
            }
        })
    }
</script>
</html>
```

在 JSP 页面中引入了 CSS 及 JavaScript 静态资源文件，Spring Boot 会默认将静态资源文件当作请求路径进行处理，所以需要对静态资源文件进行配置。在 application.properties 文件中编写配置代码，如下所示。

```
##静态资源文件配置
spring.mvc.static-path-pattern=/static/css/**
spring.mvc.static-path-pattern=/static/js/**
```

6. 测试

列表页测试结果如图 8-4 所示。

单击"修改"按钮，跳转到修改页面，如图 8-5 所示。

图 8-4　列表页测试结果　　　　　　　　图 8-5　修改页面

缓存对服务器性能的提升有非常好的效果。实现缓存的方式多种多样，本章主要介绍了 Ehcache 与 Redis 实现缓存的操作。Ehcache 不需要安装，直接在 JVM 中缓存，操作方便、快捷，但具有一定的局限性；Redis 则需要安装服务端，Java 程序作为客户端，其操作比 Ehcache 麻烦，但是 Redis 作为集群操作更为方便。

V8-2　Redis 缓存应用

第9章 项目实战

本章导读

■通过前面章节的学习，相信读者能在项目中使用相应的技术来解决具体的问题，本章将结合前面所学知识进行项目实战学习。本章将以系统中的权限管理作为示例，涉及用户、角色、权限及权限控制等功能，并讲解如何使用 Spring Boot 来快速搭建开发环境，如何通过 Spring Data 来简化数据持久层开发，以及如何通过 Shiro 来实现权限的控制。本章中涉及界面的知识相对较少，主要讨论后台业务如何实现，对前端实现有兴趣的读者可以参照蜗牛学院出版的《Web 前端开发实战教程（HTML5+CSS3+JavaScript）（微课版）》。

学习目标

（1）掌握Spring Boot、Shiro和Spring Data的整合开发。
（2）掌握集成开源项目的方法。

第 9 章 项目实战

9.1 项目介绍

本项目主要涉及 Web 开发中最基本、最核心的功能模块，包括用户管理、角色管理、权限分配等功能模块。后台采用了 Spring Boot、Spring Data、Spring MVC、Shiro，并使用 Redis 对权限控制的数据进行缓存处理，使用 FreeMarker 作为模板引擎；前端采用了 Bootstrap、jQuery 插件，并引入 FreeMarker，针对 Shiro 扩展的标签实现细粒度的权限控制，项目最终效果如图 9-1 所示。

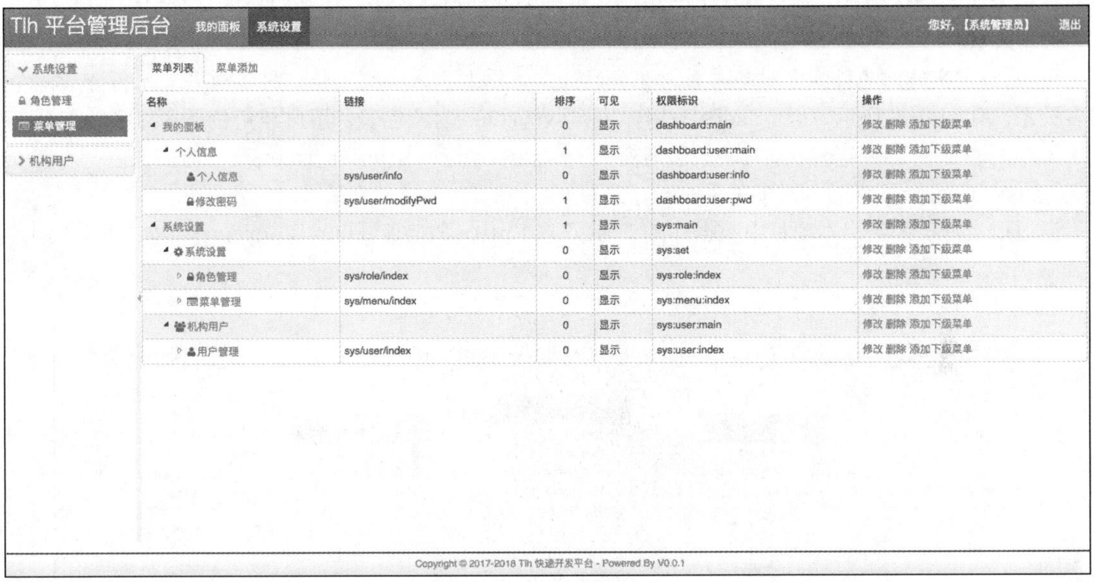

图 9-1　项目最终效果

9.2 实战开发

9.2.1 数据库设计

通过对系统中的实体对象进行分析可知，整个系统中包含的实体对象主要包括用户、角色和资源菜单，且它们的关系都是多对多，如图 9-2 所示。

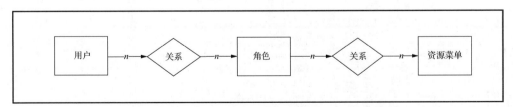

图 9-2　实体对象的关系

下面来分析每个对象中存在的属性。

1. 用户

用户主要存储登录信息和用于 Shiro 加密处理的"盐"。考虑到一般需要对用户密码进行加密，而常用的加密方式有 MD5 或 SHA-1，虽然 MD5 属于非对称加密（即使得到加密之后的数据，也无法解密得到明文信息），但是网络中仍存在破解 MD5 加密数据的技术，因此在系统设计时，除了密码字段以外，还增加了"盐"字段，以提高密码的安全性。

2．角色

角色可以看作资源菜单的集合，一个角色可以拥有多个资源菜单，该实体对象主要用于存储角色的描述信息。

3．资源菜单

资源菜单可以理解为用户可以访问的资源，系统中的每个资源都对应一个菜单，该实体对象用于存储对这些资源的描述信息。其在设计上会和自身进行关联，存在该子菜单所对应的父菜单信息。

4．关系

以上 3 个实体对象都为多对多关系，在表设计中，需要引入中间表将这种多对多关系拆分成两个一对多关系。如图 9-3 所示，sys_users 为用户表，sys_roles 为角色表，用户与角色是多对多的关系，通过引入中间表 sys_users_roles，关联关系则变为 sys_users 与 sys_users_roles 是一对多关系，sys_roles 与 sys_users_roles 同样为一对多关系。

通过以上分析，给出本项目中数据库的 E-R 图（Entity Relationship Diagram，实体-联系图），如图 9-3 所示。

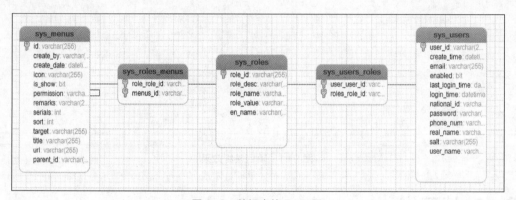

图 9-3　数据库的 E-R 图

具体的 SQL 语句如下。读者可以在本书提供的源码的 doc 文件夹中查看相应的 SQL 文件 (tlh_schema.sql)，该 SQL 文件中没有创建数据库的语句，请读者自行创建名称为 tlh 的数据库，字符编码设置为 UTF-8。

```
DROP TABLE IF EXISTS 'sys_menus';
CREATE TABLE 'sys_menus' (
 'id' varchar(255) NOT NULL,
 'create_by' varchar(255) DEFAULT NULL,
 'create_date' datetime DEFAULT NULL,
 'icon' varchar(255) DEFAULT NULL,
 'is_show' bit(1) DEFAULT NULL,
 'permission' varchar(255) DEFAULT NULL,
 'remarks' varchar(255) DEFAULT NULL,
 'serials' int(11) DEFAULT NULL,
 'sort' int(11) DEFAULT NULL,
 'target' varchar(255) DEFAULT NULL,
 'title' varchar(255) DEFAULT NULL,
 'url' varchar(255) DEFAULT NULL,
 'parent_id' varchar(255) DEFAULT NULL,
 PRIMARY KEY ('id'),
 KEY 'FKqvahr9k2gjbgu5x0hqo665ifv' ('parent_id'),
```

```sql
  CONSTRAINT 'FKqvahr9k2gjbgu5x0hqo665ifv'
    FOREIGN KEY ('parent_id') REFERENCES 'sys_menus' ('id')
) ENGINE=InnoDB DEFAULT CHARSET=utf8;

DROP TABLE IF EXISTS 'sys_roles';
CREATE TABLE 'sys_roles' (
  'role_id' varchar(255) NOT NULL,
  'role_desc' varchar(255) DEFAULT NULL,
  'role_name' varchar(255) DEFAULT NULL,
  'role_value' varchar(255) DEFAULT NULL,
  'en_name' varchar(255) DEFAULT NULL,
  PRIMARY KEY ('role_id'),
  UNIQUE KEY 'UK_3i4uf5mcxfn2mve9obk57o2u2' ('role_name'),
  UNIQUE KEY 'UK_7ui4u3k5gcop83a1ft5gq4d4o' ('en_name')
) ENGINE=InnoDB DEFAULT CHARSET=utf8;

DROP TABLE IF EXISTS 'sys_roles_menus';
CREATE TABLE 'sys_roles_menus' (
  'role_role_id' varchar(255) NOT NULL,
  'menus_id' varchar(255) NOT NULL,
  PRIMARY KEY ('role_role_id','menus_id'),
  KEY 'FK6w984a6s9uofvolb62fgjygvc' ('menus_id'),
  CONSTRAINT 'FK4o54spmf9ijamwkmnsk1dqgyb'
      FOREIGN KEY ('role_role_id') REFERENCES 'sys_roles' ('role_id'),
  CONSTRAINT 'FK6w984a6s9uofvolb62fgjygvc' FOREIGN KEY ('menus_id') REFERENCES 'sys_menus' ('id')
) ENGINE=InnoDB DEFAULT CHARSET=utf8;

DROP TABLE IF EXISTS 'sys_users';
CREATE TABLE 'sys_users' (
  'user_id' varchar(255) NOT NULL,
  'create_time' datetime DEFAULT NULL,
  'email' varchar(255) DEFAULT NULL,
  'enabled' bit(1) DEFAULT NULL,
  'last_login_time' datetime DEFAULT NULL,
  'login_time' datetime DEFAULT NULL,
  'national_id' varchar(255) DEFAULT NULL,
  'password' varchar(255) DEFAULT NULL,
  'phone_num' varchar(255) DEFAULT NULL,
  'real_name' varchar(255) DEFAULT NULL,
  'salt' varchar(255) DEFAULT NULL,
  'user_name' varchar(255) DEFAULT NULL,
  PRIMARY KEY ('user_id'),
  UNIQUE KEY 'UK_4hfk66d9netbffi5dgli7o3jq' ('user_name')
) ENGINE=InnoDB DEFAULT CHARSET=utf8;

DROP TABLE IF EXISTS 'sys_users_roles';
CREATE TABLE 'sys_users_roles' (
  'user_user_id' varchar(255) NOT NULL,
  'roles_role_id' varchar(255) NOT NULL,
  PRIMARY KEY ('user_user_id','roles_role_id'),
  KEY 'FKoplp1my1ovsuu2dnhn5c7oh0w' ('roles_role_id'),
  CONSTRAINT 'FK15wx93a8dvs2k0pqck6uqyk8h'
      FOREIGN KEY ('user_user_id') REFERENCES 'sys_users' ('user_id'),
  CONSTRAINT 'FKoplp1my1ovsuu2dnhn5c7oh0w'
```

```
    FOREIGN KEY ('roles_role_id') REFERENCES 'sys_roles' ('role_id')
) ENGINE=InnoDB DEFAULT CHARSET=utf8;
```

9.2.2　环境搭建

考虑到读者对 Eclipse 已比较熟悉，而目前企业多使用 IDEA 开发工具，且其对 Spring Boot 的支持比较好，故这里采用 IDEA 作为开发工具。

1．下载安装

IDEA 和 Eclipse 都提供了解压版和安装版，且 IDEA 提供了社区版和非社区版的安装包，相对社区版而言，非社区版的功能更加全面。这里以非社区版的安装版为例进行安装配置，读者可以自行下载安装。

2．基本配置

安装完成后，进入 IDEA 主界面，选择"Configuration"→"Preference"选项，弹出"Default Preferences"对话框，这里主要对字体和快捷键的配置做简单说明。

（1）字体设置如图 9-4 所示。

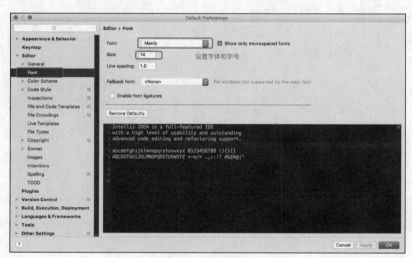

图 9-4　字体设置

（2）快捷键设置如图 9-5 所示。

图 9-5　快捷键设置

3. 配置 Maven

这里主要通过 Maven 对整个项目进行构建，所以在创建项目之前需要安装 Maven 并在 IDEA 中进行配置。关于 Maven 的安装在前面的章节中已经有所涉及，这里不再强调。配置 Maven，如图 9-6 所示。

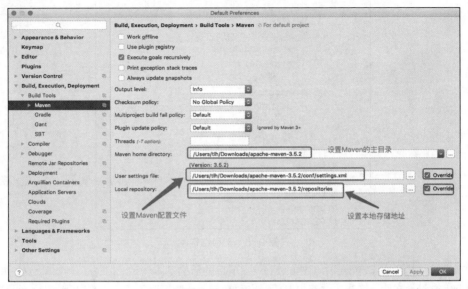

图 9-6　配置 Maven

4. 创建项目

（1）在主界面中单击"Create New Project"按钮，弹出"New Project"对话框，设置项目类型为 Maven，单击"Next"按钮，设置项目名称，如图 9-7 所示。

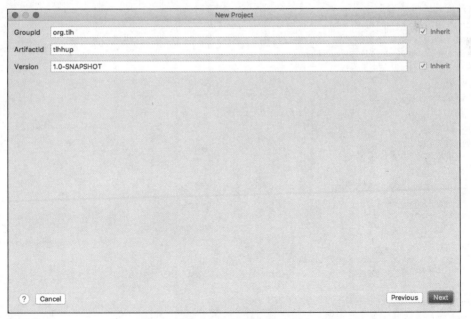

图 9-7　设置项目名称

（2）项目结构如图 9-8 所示。

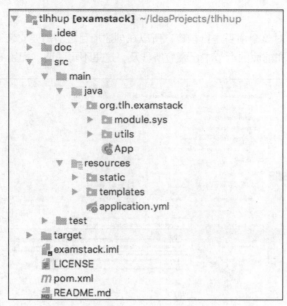

图 9-8　项目结构

整个项目的结构和在 Eclipse 中创建的 Maven 项目的结构基本一致，但多了一个.idea 文件夹和一个扩展名为.iml 的配置文件，这些文件都由 IDEA 自动维护，一般情况下不需要做任何改动。同时，项目包结构和资源目录的结构也如图 9-8 所示，对此不再赘述。

5. 添加依赖

修改 pom.xml 文件，添加 Spring MVC、Spring Data JPA、FreeMarker 和数据库相关的依赖信息，引入 HikariCP 数据源，代码如下。

```xml
<parent>
    <groupId>org.springframework.boot</groupId>
    <artifactId>spring-boot-starter-parent</artifactId>
    <version>1.4.2.RELEASE</version>
</parent>
<properties>
    <shiro.version>1.4.0</shiro.version>
    <hikariCP.version>2.7.1</hikariCP.version>
</properties>
<dependencies>
    <!-- 集成Spring MVC -->
    <dependency>
        <groupId>org.springframework.boot</groupId>
        <artifactId>spring-boot-starter-web</artifactId>
    </dependency>
    <!-- 集成Spring Data -->
    <dependency>
        <groupId>org.springframework.boot</groupId>
        <artifactId>spring-boot-starter-data-jpa</artifactId>
    </dependency>
    <!-- 集成FreeMarker -->
    <dependency>
        <groupId>org.springframework.boot</groupId>
        <artifactId>spring-boot-starter-freemarker</artifactId>
    </dependency>
    <!-- MySQL -->
```

```xml
        <dependency>
            <groupId>mysql</groupId>
            <artifactId>mysql-connector-java</artifactId>
        </dependency>
        <!-- Hikari -->
        <dependency>
            <groupId>com.zaxxer</groupId>
            <artifactId>HikariCP</artifactId>
            <version>${hikariCP.version}</version>
        </dependency>
        <!-- Servlet -->
        <dependency>
            <groupId>javax.servlet</groupId>
            <artifactId>javax.servlet-api</artifactId>
            <version>3.1.0</version>
        </dependency>
        <!-- test -->
        <dependency>
            <groupId>org.springframework.boot</groupId>
            <artifactId>spring-boot-starter-test</artifactId>
        </dependency>
</dependencies>
```

6. 编写主类并进行测试

（1）编写配置文件。通过前面章节的学习，可以了解到 Spring Boot 支持两种格式的配置文件，即 properties 格式和 yml 格式，关于属性格式的配置文件读者已经比较熟悉了，这里将采用第二种配置文件格式。在 resources 目录中创建 application.yml 文件，配置日志、数据源、JPA 和 FreeMarker 的相关信息，具体内容如下。

```yaml
logging:
  level: debug
spring:
  datasource:
    driver-class-name: org.gjt.mm.mysql.Driver
    url: jdbc:mysql:///tlh?useUnicode=true&characterEncoding=utf-8
    username: admin
    password: 123456
    type: com.zaxxer.hikari.HikariDataSource
  jpa:
    generate-ddl: true
    show-sql: true
  freemarker:
    charset: utf-8
    cache: false
    enabled: true
server:
  port: 8089
  context-path: /tlh
```

（2）编写启动类。在项目最顶层的包中创建 App 类并将其作为启动类，其内容和前面学习 Spring Boot 时编写的启动类内容一致，唯一的区别就是这里需要添加开启注解事务和指定 JPA 的 Repository 存储的包路径，具体代码如下。

```java
@SpringBootApplication
@EnableTransactionManagement
@EnableJpaRepositories(basePackages = "org.tlh.examstack.module.**.repository")
public class App {
```

```
    public static void main(String[] args){
        SpringApplication.run(App.class,args);
    }
}
```

9.2.3 用户管理

将项目源码中 static 文件中的静态资源复制到项目 resources 目录的 static 文件夹中，在 templates 文件夹中创建 sys 文件夹，用于存储系统相关的管理界面。

1．编写用户实体类

编写用户实体类对象时，为方便后面通过 JPA 来完成数据的持久化操作，将主键生成策略设置为 UUID，这里不设置为自增主键是因为整型数据在后期用户量增多之后，进行分库分表时可能会出现诸如主键重复等问题，或者在导入数据库的时候，可能会因为主键而出现一些问题。具体编写代码如下。

```
@Entity
@Table(name = "sys_users")
public class User extends BaseEntity {
    @Id
    @GenericGenerator(strategy = "uuid", name = "uuid")
    @GeneratedValue(generator = "uuid")
    private String userId;

    @Column(unique = true)
    private String userName;
    private String password;
    private String realName;
    private String nationalId;
    private String email;
    private String phoneNum;
    private Date createTime;
    private Date lastLoginTime;
    private Date loginTime;
    private Boolean enabled;
    private String salt;
}
```

查看该用户实体类的属性，相信读者知道其中大部分属性所表示的含义，这里只针对其中一个比较特殊的字段"salt"进行说明。"salt"可以理解为调味料"盐"，但这里也可以作如下理解。对于用户信息而言，一般需要对密码进行加密处理，而常用的方式为先采用 MD5 或 SHA-1 算法得到明文的数字摘要，再通过 Base64 等进行编码得到加密之后的密文，但这样处理过于简单，即使是这里可以进行多次加密，但加密算法和处理方式都是透明的，所以仍然不安全。这时，"salt"就应运而生了，它如同一味调味料，对密码进行混淆。"salt"的产生由系统进行处理，外部是无法知道其是如何计算获取的，故而提高了密码的安全性。

2．编写用户的存储接口

这里借助 Spring Data JPA 中提供的基类作为系统中持久层模板的基类，处理增、删、改、查等基本操作，关于 JpaRepository 中提供的基础方法可以回顾第 4 章中 Spring Data 的相关知识点，存储接口代码如下。

```
public interface UserRepository extends JpaRepository<User,String>{

    User findByUserName(String userName);
```

```
    List<User> findUserByRoles(Role role);

    @Query("update User u set u.realName=:#{#user.realName},
            u.email=:#{#user.email},u.phoneNum=:#{#user.phoneNum} ,
            u.enabled=:#{#user.enabled} where u.userId=:#{#user.userId}")
    @Modifying
    @Transactional
    int updateUserInfo(@Param("user") User user);

    @Query("update User u set u.realName=:#{#user.realName},
            u.email=:#{#user.email}, u.phoneNum=:#{#user.phoneNum} ,
            u.enabled=:#{#user.enabled}, u.salt=:#{#user.salt},
            u.password=:#{#user.password} where  u.userId=:#{#user.userId}")
    @Modifying
    @Transactional
    int updateUserInfoWithPwd(@Param("user") User user);

    @Query("update User u set u.password=?1 where u.userId=?2")
    @Modifying
    @Transactional
    int modifyPwd(String password,String id);

    /**
     * 更新用户最后登录的时间
     * @param user
     */
    @Query("update User u set u.lastLoginTime=:#{#user.lastLoginTime}
            where u.userId=:#{#user.userId}")
    @Modifying
    @Transactional
    void updateLastLogin(@Param("user") User user);
}
```

针对该接口中的参数绑定进行补充：在第 4 章中介绍 Spring Data JPA 的时候，参数绑定主要是通过占位符的方式实现的，但是这在某些场景下并不太合适，如更新用户信息时，需要为每个字段都添加相应的参数以进行数据绑定，由于涉及的字段太多，过于麻烦，此时可以使用命名参数的方式实现数据绑定，如前面接口中的 updateUserInfo 方法，代码如下。

```
@Query("update User u set u.realName=:#{#user.realName},
        u.email=:#{#user.email},u.phoneNum=:#{#user.phoneNum},
        u.enabled=:#{#user.enabled} where u.userId=:#{#user.userId}")
@Modifying
@Transactional
int updateUserInfo(@Param("user") User user);
```

以上采用 ":#{命名参数}" 的语法实现了参数的定义。其中，#user.realName 表示将变量 user 的 realName 属性的值绑定到该命名参数中，参数列表中的@Param注解表示传递到 JPQL 语句中的参数名称。

3．编写用户业务接口

针对业务接口，最基本的操作是添加、删除、修改以及通过主键查询，这里先通过提取基类对这些业务方法进行统一处理，在 Service 包中创建 BaseService 接口（泛型接口），代码如下。

```
public interface BaseService<T,ID extends Serializable> {

    T saveOrUpdate(T t);

    void delete(ID id);
```

```
    T getOne(ID id);

    boolean exists(Example<T> example);

    Page<T> findWithPageInfo(Example<T> example,Pageable pageable);

    Page<T> findAll(Pageable pageable);

    List<T> findAll();
}
```

在 impl 包中编写该接口的实现类 BaseServiceImpl，代码如下。

```
@Transactional(readOnly = true)
public abstract class BaseServiceImpl<T,ID extends Serializable> implements BaseService<T,ID> {

    private JpaRepository<T,ID> mJpaRepository;

    public void setJpaRepository(JpaRepository<T, ID> jpaRepository) {
        this.mJpaRepository = jpaRepository;
    }

    @Transactional(readOnly = false)
    @Override
    public T saveOrUpdate(T t) {
        return this.mJpaRepository.save(t);
    }

    @Transactional(readOnly = false)
    @Override
    public void delete(ID id) {
        this.mJpaRepository.delete(id);
    }

    @Override
    public T getOne(ID id) {
        return this.mJpaRepository.findOne(id);
    }

    @Override
    public boolean exists(Example<T> example) {
        return this.mJpaRepository.exists(example);
    }

    @Override
    public Page<T> findWithPageInfo(Example<T> example, Pageable pageable) {
        return this.mJpaRepository.findAll(example,pageable);
    }

    @Override
    public Page<T> findAll(Pageable pageable) {
        return this.mJpaRepository.findAll(pageable);
    }

    @Override
```

```
    public List<T> findAll() {
        return this.mJpaRepository.findAll();
    }
}
```

该实现类通过 Spring 的 IoC 机制将 Repository 层的实现注入进来。注意，这里注入的类型为 JpaRepository，因为它是 Repository 层的基类，因此能保证基础业务接口的通用性。

这样就完成了对服务层基类的抽取，但需要注意的是，如果没有 JpaRepository 的存在（准确来说是 Repository 层基类的存在），是无法抽取服务层的依赖的。也可以自己实现一个具有 JpaRepository 功能的基类，有兴趣的读者可以自行实现。继续编写用户的服务接口，在 Service 包中添加 UserService 接口，其继承 BaseService 的同时添加自己的服务方法，具体代码如下。

```
public interface UserService extends BaseService<User,String>{

    /**
     * 通过用户名获取用户信息
     * @param userName
     * @return
     */
    User findUserByUserName(String userName);

    /**
     * 更新用户信息
     * @param user
     * @return
     */
    boolean updateUserInfo(User user);

    /**
     * 修改用户密码
     * @param user
     * @return
     */
    boolean modifyPwd(User user);

    /**
     * 检测用户输入的原始密码是否正确
     * @param oldPassword
     * @param userId
     * @return
     */
    boolean checkOldPwd(String oldPassword, String userId);

    /**
     * 查询用户的角色
     * @param userName
     * @return
     */
    List<String> findRolesByUserName(String userName);

    /**
     * 查询用户的权限
     * @param userName
     * @return
     */
    List<String> findMenusByUserName(String userName);
```

```java
    /**
     * 通过角色ID获取用户信息
     * @param role，必须对角色ID进行赋值
     * @return
     */
    List<User> findUsersByRole(Role role);

    /**
     * 更新用户最后登录的时间
     * @param user
     */
    void updateLastLogin(User user);
}
```

这里给出了 UserService 中定义的所有服务方法，关于每个方法的功能查看相应的注释即可，至于在哪里调用这些服务方法，将在后面进行详细介绍。接口定义好之后，继续在 impl 包中定义该接口的实现类，代码如下。

```java
@Service("userService")
@Transactional(readOnly = true)
public class UserServiceImpl extends BaseServiceImpl<User, String> implements UserService {

    @Resource(name = "userRepository")
    @Override
    public void setJpaRepository(JpaRepository<User, String> jpaRepository) {
        super.setJpaRepository(jpaRepository);
    }

    @Resource
    private UserRepository mUserRepository;

    @Override
    public User findUserByUserName(String userName) {
        return this.mUserRepository.findByUserName(userName);
    }

    @Transactional
    @Override
    public User saveOrUpdate(User user) {
        if (StringUtils.isEmpty(user.getUserId())) {  //新增用户
            user.setLastLoginTime(new Date());
            user.setCreateTime(new Date());
            user.setLoginTime(new Date());
            prepareUser(user);

            return this.mUserRepository.save(user);
        } else if (!checkOldPwd(user.getPassword(), user.getUserId())) {
            /* 如果修改了密码，则通过JPQL来修改用户信息，
               若通过merge来修改将会导致角色信息丢失 */
            prepareUser(user);
            mUserRepository.updateUserInfoWithPwd(user);
        } else {
            //如果没有修改密码
            this.mUserRepository.updateUserInfo(user);
        }
```

```java
        return user;
}

@Transactional
@Override
public boolean updateUserInfo(User user) {
    return this.mUserRepository.updateUserInfo(user) > 0;
}

@Override
public boolean checkOldPwd(String oldPassword, String userId) {
    User user = this.getOne(userId);
    //构建比对用户
    User oldUser = new User();
    oldUser.setSalt(user.getSalt());
    oldUser.setPassword(oldPassword);
    prepareUser(oldUser);
    //比对密码
    return oldUser.getPassword().equals(user.getPassword());
}

@Autowired
private JdbcTemplate jdbcTemplate;

@Override
public List<String> findRolesByUserName(String userName) {
    String sql = "SELECT\n" +
            "\tsr.en_name\n" +
            "FROM\n" +
            "\tsys_users su\n" +
            "INNER JOIN sys_users_roles sur ON su.user_id = sur.user_user_id\n" +
            "INNER JOIN sys_roles sr ON sur.roles_role_id = sr.role_id\n" +
            "WHERE\n" +
            "\tsu.user_name = ?";
    return this.jdbcTemplate.query(sql, new Object[]{userName},
            new SingleColumnRowMapper<String>());
}

@Override
public List<String> findMenusByUserName(String userName) {
    String sql = "SELECT\n" +
            "\tsm.permission\n" +
            "FROM\n" +
            "\tsys_users su\n" +
            "INNER JOIN sys_users_roles sur ON su.user_id = sur.user_user_id\n" +
            "INNER JOIN sys_roles sr ON sur.roles_role_id = sr.role_id\n" +
            "INNER JOIN sys_roles_menus srm on sr.role_id=srm.role_role_id\n" +
            "INNER JOIN sys_menus sm on srm.menus_id=sm.id\n" +
            "WHERE\n" +
            "\tsu.user_name = ?";
    return this.jdbcTemplate.query(sql, new Object[]{userName},
            new SingleColumnRowMapper<String>());
}

@Transactional
```

```java
    @Override
    public boolean modifyPwd(User user) {
        prepareUser(user);
        return this.mUserRepository.modifyPwd(user.getPassword(), user.getUserId()) > 0;
    }

    @Override
    public List<User> findUsersByRole(Role role) {
        return this.mUserRepository.findUserByRoles(role);
    }

    @Transactional(readOnly = false)
    @Override
    public void updateLastLogin(User user) {
        this.mUserRepository.updateLastLogin(user);
    }

    //设置盐并对密码进行加密
    private void prepareUser(User user) {
        if (StringUtils.isEmpty(user.getSalt())) {
            //生成凭证的盐,加强密码的安全性
            String salt = new SecureRandomNumberGenerator().nextBytes().toHex();
            salt = user.getUserName() + salt;
            user.setSalt(salt);
        }
        //加密
        SimpleHash hash = new SimpleHash(Md5Hash.ALGORITHM_NAME,
                    user.getPassword(), user.getSalt(), 5);
        user.setPassword(hash.toHex());
    }
}
```

在该实现类中注入了 UserRepository 和 JdbcTemplate 来实现对数据库的操作,之所以引入 JdbcTemplate 是考虑到在某些场景中直接通过 JDBC 来操作更为方便。该实现类中除了 UserService 中声明的方法以外,还有一个名为 prepareUser 的方法,这个方法主要用于实现加盐处理,以加强密码的安全性,通过盐和 5 次加密处理后,密码的安全性提升。

4. 编写处理器

服务层代码编写完成之后,要处理和界面的交互,接收并处理用户请求。在 controller 包中编写 UserController,代码如下。

```java
@Controller
@RequestMapping("/sys/user")
public class UserController {

    @Autowired
    private UserService userService;

    @RequestMapping(value = "/info",method = RequestMethod.GET)
    public String info(Model model){
        model.addAttribute("user",
                SecurityUtils.getSubject().getPrincipals().oneByType(User.class));
        return "sys/user/info";
    }
```

```java
    @RequestMapping(value = "/info",method = RequestMethod.POST)
    public String info(User user,Model model){
        if(this.userService.updateUserInfo(user)){
            model.addAttribute("user",this.userService.getOne(user.getUserId()));
            model.addAttribute("message","修改成功");
        }else {
            model.addAttribute("message","修改失败");
            model.addAttribute("user",user);
        }
        return "sys/user/info";
    }

    @RequestMapping(value = "/modifyPwd",method = RequestMethod.GET)
    public String modifyPwd(Model model){
        model.addAttribute("userId",
            SecurityUtils.getSubject().getPrincipals().oneByType(User.class).getUserId());
        return "sys/user/modifyPwd";
    }

    @RequestMapping(value = "/modifyPwd",method = RequestMethod.POST)
    public String modifyPwd(
        String oldPassword,String newPassword,String userId,Model model){
        if(this.userService.checkOldPwd(oldPassword,userId)){
            User user = SecurityUtils.getSubject().getPrincipals().oneByType(User.class);
            user.setPassword(newPassword);
            this.userService.modifyPwd(user);
        }else{
            model.addAttribute("message","修改密码失败,旧密码错误");
        }
        model.addAttribute("userId",userId);
        return "sys/user/modifyPwd";
    }

    @RequestMapping(value = "/index",method = {RequestMethod.GET,RequestMethod.POST})
    public String index(
        User user, @PageableDefault(value = 10) Pageable pageable, Model model){
        if(StringUtils.isEmpty(user.getUserName())){
            user.setUserName(null);
        }
        if(StringUtils.isEmpty(user.getRealName())){
            user.setRealName(null);
        }
        ExampleMatcher exampleMatcher =
        ExampleMatcher.matching().
        withMatcher("userName", matcher -> matcher.startsWith()).
        withMatcher("realName", matcher -> matcher.startsWith());
        Example<User> example=Example.of(user,exampleMatcher);
        model.addAttribute("users",this.userService.findWithPageInfo(example,pageable));
        return "sys/user/userList";
    }

    @RequestMapping(value = "/form",method = RequestMethod.GET)
```

```java
    public String form(String id,Model model){
        User user=null;
        if(StringUtils.isEmpty(id)){
            user=new User();
        }else {
            user=this.userService.getOne(id);
        }
        model.addAttribute("user",user);
        return "sys/user/userFrom";
    }

    @RequestMapping(value = "/checkLoginName",method = RequestMethod.GET)
    public @ResponseBody boolean checkLoginName(String userName){
        return this.userService.findUserByUserName(userName)==null;
    }

    @RequestMapping(value = "/save",method = RequestMethod.POST)
    public String save(User user){
        if(StringUtils.isEmpty(user.getPassword())){   //更新用户信息
            this.userService.updateUserInfo(user);
        }else{
            user=this.userService.saveOrUpdate(user);
        }
        return "redirect:form?id="+user.getUserId();
    }

    @RequestMapping(value = "/delete",method = RequestMethod.GET)
    public String delete(String id){
        this.userService.delete(id);
        return "redirect:index";
    }
}
```

这里给出了所有方法的声明，每个方法表示的含义通过方法名即可得出，这里不再一一介绍。需要强调的是，针对重载的方法，有的接收请求方法为 GET，而有的为 POST，它们是程序开发中比较常见的方式，GET 方法用于界面展示，而 POST 方法用于业务处理。除此之外，在 index 方法中创建的 Matcher 对象使用了 JDK 1.8 中的 Lambda 表达式，关于这方面的知识，读者可自行查找学习，这里不再赘述。

5. 编写列表界面

在 templates 文件中创建 sys/user 文件夹，并在该文件夹中创建 userList.ftl 文件，代码如下。

```html
<html>
<head>
    <title>用户管理</title>
    <meta name="decorator" content="default"/>
    <#include "../../layout/header.ftl">
    <style>
        #contentTable{
            font-size: 13px;
        }
    </style>
</head>
<body>
<div id="importBox" class="hide">
    <form id="importForm" action="${ctx}/sys/user/import" method="post"
          enctype="multipart/form-data"class="form-search">
```

```html
                    style="padding-left:20px;text-align:center;"
                    onsubmit="loading('正在导入，请稍等...');"><br/>
            <input id="uploadFile" name="file" type="file"
                    style="width:330px"/><br/><br/>
            <input id="btnImportSubmit"
                    class="btn btn-primary" type="submit" value="   导     入   "/>
            <a href="${ctx}/sys/user/import/template">下载模板</a>
    </form>
</div>
<ul class="nav nav-tabs">
    <li class="active"><a href="${ctx}/sys/user/index">用户列表</a></li>
    <@shiro.hasPermission name="sys:user:form">
        <li><a href="${ctx}/sys/user/form">用户添加</a></li>
    </@shiro.hasPermission>
</ul>
<form id="searchForm" modelAttribute="user"
            action="${ctx}/sys/user/index" method="post" class="breadcrumb form-search">
    <input id="pageNo" name="page" type="hidden" value="${users.number}"/>
    <input id="pageSize" name="size" type="hidden" value="${users.size}"/>
    <ul class="ul-form">
        <li><label>登录名：</label>
            <input name="userName" htmlEscape="false" maxlength="50"
                class="input-medium"/></li>
        <li><label>姓   名：</label>
            <input name="realName" htmlEscape="false" maxlength="50"
                class="input-medium"/></li>
        <li class="btns">
            <input id="btnSubmit" class="btn btn-primary" type="submit" value="查询"
                onclick="return page();"/>
            <input id="btnExport" class="btn btn-primary" type="button" value="导出"/>
            <input id="btnImport" class="btn btn-primary" type="button" value="导入"/></li>
        <li class="clearfix"></li>
    </ul>
</form>
<table id="contentTable" class="table table-striped table-bordered table-condensed">
    <thead><tr>
            <th class="sort-column login_name">登录名</th>
            <th class="sort-column name">姓名</th>
            <th>电话</th>
            <th>邮箱</th>
            <th>创建日期</th>
            <th>最后登录</th>
            <th>是否可用</th>
            <@shiro.hasPermission name="sys:user:edit">
            <th>操作</th></@shiro.hasPermission></tr></thead>
    <tbody>
    <#list users.content as user>
        <tr>
            <td><a    href="${ctx}/sys/user/form?id=${user.userId}">${user.userName}</a></td>
            <td>${user.realName}</td>
            <td>${user.phoneNum}</td>
            <td>${user.email}</td>
            <td>${user.createTime}</td>
```

```
                    <td>${user.lastLoginTime}</td>
                    <td>${user.enabled?then('可用','禁用')}</td>
                    <@shiro.hasPermission name="sys:user:edit">
                        <td>
                            <a href="${ctx}/sys/user/form?id=${user.userId}">修改</a>
                            <a href="${ctx}/sys/user/delete?id=${user.userId}"
                                onclick="return confirmx('确认要删除该用户吗？', this.href)">
                                删除
                            </a>
                        </td>
                    </@shiro.hasPermission>
                </tr>
        </#list>
        </tbody>
</table>

<div class="pagination">
    <ul>
        <#if users.first>
            <li class="disabled"><a href="javascript:">« 上一页</a></li>
        <#else>
            <li>
              <a href="javascript:"
                    onclick="page(${users.number-1},${users.size},'');">
                    « 上一页</a></li>
        </#if>
        <#assign pages=(users.totalPages>3)?then(3,users.totalPages)>
        <#list 0..<pages as pageNumber>
            <li class="${((users.number)==pageNumber)?then('active','')}">
            <a href="javascript:" onclick="page(${pageNumber},${users.size},'');">
            ${pageNumber+1}</a></li>
        </#list>
        <#if users.last>
            <li class="disabled"><a href="javascript:">下一页 »</a></li>
        <#else>
            <li>
              <a href="javascript:" onclick="page(${users.number+1},${users.
                size},'');">下一页 »</a></li>
        </#if>
        <li class="disabled controls">
            <a href="javascript:">
                当前
                <input type="text" value="${users.number+1}"
                    onkeypress="var e=window.event||event;
                    var c=e.keyCode||e.which;
                    if(c==13)page(${users.number},${users.size},'');"
                      onclick="this.select();"> /
                <input type="text" value="${users.size}"
                    onkeypress="var e=window.event||event;
                    var c=e.keyCode||e.which;if(c==13)page(0,this.value,'');"
                      onclick="this.select();">
                条，共 ${users.totalElements} 条
            </a>
        </li>
    </ul>
```

```
        <div style="clear:both;"></div>
</div>
</body>
<script type="text/javascript">
    $(document).ready(function() {
        $("#btnExport").click(function(){
            top.$.jBox.confirm("确认要导出用户数据吗?","系统提示",function(v,h,f){
                if(v=="ok"){
                    $("#searchForm").attr("action","${ctx}/sys/user/export");
                    $("#searchForm").submit();
                }
            },{buttonsFocus:1});
            top.$('.jbox-body .jbox-icon').css('top','55px');
        });
        $("#btnImport").click(function(){
            $.jBox($("#importBox").html(), {title:"导入数据", buttons:{"关闭":true},
                bottomText:"导入文件不能超过5MB,仅允许导入\"xls\"或\"xlsx\"格式文件!"});
        });
    });
    function page(n,s){
        if(n>=0) $("#pageNo").val(n);
        if(s) $("#pageSize").val(s);
        $("#searchForm").attr("action","${ctx}/sys/user/index");
        $("#searchForm").submit();
        return false;
    }
</script>
</html>
```

在该界面中使用了 Shiro 中的部分标签,这部分内容在讲解权限控制时会进行详细说明,这里可以暂不探究。

6. 编写修改密码界面

在上面创建的文件夹中创建 modifyPwd.ftl 文件,通过调用 UserController 中的 modifyPwd 方法实现对密码的修改,代码如下。

```
<html>
<head>
    <title>修改密码</title>
    <meta name="decorator" content="default"/>
    <#include "../../layout/header.ftl">
    <script type="text/javascript">
        $(document).ready(function() {
            $("#oldPassword").focus();
            $("#inputForm").validate({
                rules: {
                },
                messages: {
                    confirmNewPassword: {equalTo: "输入与上面相同的密码"}
                },
                submitHandler: function(form){
                    loading('正在提交,请稍等...');
                    form.submit();
                },
                errorContainer: "#messageBox",
                errorPlacement: function(error, element) {
                    $("#messageBox").text("输入有误,请先更正。");
```

```
                        if (element.is(":checkbox")||element.is(":radio")
                            ||element.parent().is(".input-append")){
                            error.appendTo(element.parent().parent());
                        } else {
                            error.insertAfter(element);
                        }
                    }
                });
            });
        </script>
    </head>
    <body>
    <ul class="nav nav-tabs">
        <li><a href="${ctx}/sys/user/info">个人信息</a></li>
        <li class="active"><a href="${ctx}/sys/user/modifyPwd">修改密码</a></li>
    </ul><br/>
    <form id="inputForm" action="${ctx}/sys/user/modifyPwd"
                        method="post" class="form-horizontal">
        <input type="hidden" name="userId" value="${userId}">
        <#if message??>
            <div id="messageBox" class="alert alert-error">
                <button data-dismiss="alert" class="close">×</button>
                ${message}
            </div>
        </#if>
        <#-- 关闭错误提示 -->
        <script type="text/javascript">top.$.jBox.closeTip();</script>
        <div class="control-group">
            <label class="control-label">旧密码:</label>
            <div class="controls">
                <input id="oldPassword" name="oldPassword" type="password" value=""
                                        maxlength="50" minlength="3" class="required"/>
                <span class="help-inline"><font color="red">*</font> </span>
            </div>
        </div>
        <div class="control-group">
            <label class="control-label">新密码:</label>
            <div class="controls">
                <input id="newPassword" name="newPassword" type="password"
                            value="" maxlength="50" minlength="3" class="required"/>
                <span class="help-inline"><font color="red">*</font> </span>
            </div>
        </div>
        <div class="control-group">
            <label class="control-label">确认新密码:</label>
            <div class="controls">
                <input id="confirmNewPassword" name="confirmNewPassword"
                            type="password" value="" maxlength="50" minlength="3"
                                class="required" equalTo="#newPassword"/>
                <span class="help-inline"><font color="red">*</font> </span>
            </div>
        </div>
        <div class="form-actions">
            <input id="btnSubmit" class="btn btn-primary" type="submit" value="保 存"/>
        </div>
```

```
</form>
</body>
</html>
```

7. 编写用户添加、更新界面

创建 userForm.ftl 文件,该文件用于添加用户信息。通过调用 UserController 中的 form 方法实现添加或更新操作,二者主要通过 ID 进行区分,如果 ID 存在,则表示更新用户信息,否则表示添加用户信息。具体代码如下。

```
<html>
<head>
    <title>用户管理</title>
    <meta name="decorator" content="default"/>
    <#include "../../layout/header.ftl">
</head>
<body>
<ul class="nav nav-tabs">
    <li><a href="${ctx}/sys/user/index">用户列表</a></li>
    <li class="active"><a href="${ctx}/sys/user/form?id=${user.userId!}">用户
        <@shiro.hasPermission name="sys:user:edit">
        ${(user.userId??)?then('修改','添加')}
        </@shiro.hasPermission><@shiro.lacksPermission name="sys:user:edit">查看
        </@shiro.lacksPermission></a></li>
</ul><br/>
<form   id="inputForm"   action="${ctx}/sys/user/save"   method="post"   class="form-
horizontal">
    <script type="text/javascript">top.$.jBox.closeTip();</script>
    <input type="hidden" name="userId" value="${user.userId!}">
    <div class="control-group">
        <label class="control-label">登录名:</label>
        <div class="controls">
            <input name="userName" htmlEscape="false" maxlength="50" class="required
                userName"    ${(user.userId??)?then('readonly','')}    value="${user.
userName!}"/>
            <span class="help-inline"><font color="red">*</font> </span>
        </div>
    </div>
    <div class="control-group">
        <label class="control-label">姓名:</label>
        <div class="controls">
            <input name="realName" htmlEscape="false" maxlength="50" class="required"
                value="${user.realName!}"/>
            <span class="help-inline"><font color="red">*</font> </span>
        </div>
    </div>
    <div class="control-group">
        <label class="control-label">密码:</label>
        <div class="controls">
            <input id="newPassword" name="password" type="password"
                value="${user.password!}" maxlength="50" minlength="3"
                class="${(!user.userId??)?then('required','')}"/>
            <#if !user.userId??><span class="help-inline"><font color="red">*</font>
                </span></#if>
            <#if user.userId??><span class="help-inline">若不修改密码,请留空。</span></#if>
        </div>
    </div>
    <div class="control-group">
        <label class="control-label">确认密码:</label>
```

```
                <div class="controls">
                    <input id="confirmNewPassword" name="confirmNewPassword" type="password"
                        value="${user.password!}" maxlength="50" minlength="3"
                        equalTo="#newPassword"/>
                    <#if !user.userId??>
                        <span class="help-inline"><font color="red">*</font> </span></#if>
                </div>
        </div>
        <div class="control-group">
            <label class="control-label">邮箱:</label>
            <div class="controls">
                <input name="email" htmlEscape="false" maxlength="100" class="email"
                    value="${user.email!}"/>
            </div>
        </div>
        <div class="control-group">
            <label class="control-label">电话:</label>
            <div class="controls">
                <input name="phoneNum" htmlEscape="false" maxlength="100"
                    value="${user.phoneNum!}"/>
            </div>
        </div>
        <div class="control-group">
            <label class="control-label">是否允许登录:</label>
            <div class="controls">
                <input name="enabled" type="radio"
                        htmlEscape="false" class="required" value="1"
                        ${(user.enabled??)?then(user.enabled?then('checked',''),'checked')}>是
                <input name="enabled" type="radio"
                        htmlEscape="false" class="required" value="0"
                        ${(user.enabled??)?then((!user.enabled)?then('checked',''),'')}>否
                <span class="help-inline">
                    <font color="red">*</font>
                    "是"代表此账号允许登录,"否"则表示此账号不允许登录</span>
            </div>
        </div>
        <#if user.userId??>
            <div class="control-group">
                <label class="control-label">创建时间:</label>
                <div class="controls">
                    <label class="lbl">${user.createTime!}</label>
                </div>
            </div>
            <div class="control-group">
                <label class="control-label">最后登录:</label>
                <div class="controls">
                    <label class="lbl">${user.lastLoginTime!}</label>
                </div>
            </div>
        </#if>
        <div class="form-actions">
            <@shiro.hasPermission name="sys:user:edit">
              <input id="btnSubmit" class="btn btn-primary" type="submit" value="保 存"/> 
            </@shiro.hasPermission>
            <input id="btnCancel" class="btn" type="button" value="返 回"
```

```
                onclick="history.go(-1)"/>
        </div>
</form>
</body>
<script type="text/javascript">
    $(document).ready(function() {
        $("#no").focus();
        $("#inputForm").validate({
            rules: {
                <#if !user.userId??>
                userName: {
                    remote: "${ctx}/sys/user/checkLoginName"
                }
                </#if>
            },
            messages: {
                userName: {remote: "用户登录名已存在"},
                confirmNewPassword: {equalTo: "输入与上面相同的密码"}
            },
            submitHandler: function(form){
                loading('正在提交，请稍等...');
                form.submit();
            },
            errorContainer: "#messageBox",
            errorPlacement: function(error, element) {
                $("#messageBox").text("输入有误，请先更正。");
                if (element.is(":checkbox")
                ||element.is(":radio")||element.parent().is(".input-append")){
                    error.appendTo(element.parent().parent());
                } else {
                    error.insertAfter(element);
                }
            }
        });
    });
</script>
</html>
```

8. 编写用户信息详情界面

创建 info.ftl 文件，其主要用于用户信息的展示，主要通过 UserController 中的 info 方法进行处理，通过用户 ID 获取数据库中的用户信息，代码如下。

```
<html>
<head>
    <title>个人信息</title>
    <meta name="decorator" content="default"/>
    <#include "../../layout/header.ftl">
    <script type="text/javascript">
        $(document).ready(function() {
            $("#inputForm").validate({
                submitHandler: function(form){
                    loading('正在提交，请稍等...');
                    form.submit();
                },
                errorContainer: "#messageBox",
                errorPlacement: function(error, element) {
                    $("#messageBox").text("输入有误，请先更正。");
```

```
                    if (element.is(":checkbox")
                        ||element.is(":radio")
                        ||element.parent().is(".input-append")){
                            error.appendTo(element.parent().parent());
                    } else {
                        error.insertAfter(element);
                    }
                }
            });
        });
    </script>
</head>
<body>
<ul class="nav nav-tabs">
    <li class="active"><a href="${ctx}/sys/user/info">个人信息</a></li>
    <@shiro.hasPermission name="dashboard:user:pwd">
        <li><a href="${ctx}/sys/user/modifyPwd">修改密码</a></li>
    </@shiro.hasPermission>
</ul><br/>
<form id="inputForm" action="${ctx}/sys/user/info" method="post" class="form-horizontal">
    <input type="hidden" name="userId" value="${user.userId}">
    <#if message??>
        <div id="messageBox" class="alert alert-error">
            <button data-dismiss="alert" class="close">×</button>
            ${message}
        </div>
    </#if>
    <#-- 关闭错误提示 -->
    <script type="text/javascript">top.$.jBox.closeTip();</script>
    <div class="control-group">
        <label class="control-label">用户名:</label>
        <div class="controls">
            <input name="userName" htmlEscape="false" maxlength="50" class="required"
                    readonly="true" value="${user.userName!}"/>
        </div>
    </div>
    <div class="control-group">
        <label class="control-label">姓名:</label>
        <div class="controls">
            <input name="realName" htmlEscape="false" maxlength="50" class="required"
                    value="${user.realName!}"/>
        </div>
    </div>
    <div class="control-group">
        <label class="control-label">邮箱:</label>
        <div class="controls">
            <input name="email" htmlEscape="false" maxlength="50" class="email"
                    value="${user.email!}"/>
        </div>
    </div>
    <div class="control-group">
        <label class="control-label">电话:</label>
        <div class="controls">
            <input name="phoneNum" htmlEscape="false" maxlength="50"
```

```
                                value="${user.phoneNum!}"/>
            </div>
        </div>
        <div class="control-group">
            <label class="control-label">上次登录:</label>
            <div class="controls">
                <label class="lbl">时间: <#if
                                user.lastLoginTime??>${user.lastLoginTime?datetime}</label>
</#if>
            </div>
        </div>
        <div class="form-actions">
            <input id="btnSubmit" class="btn btn-primary" type="submit" value="保 存"/>
        </div>
</form>
</body>
</html>
```

通过本小节对用户管理的讲解，读者应初步掌握了整个系统的框架及其实现流程和逻辑。下面将对角色信息的维护进行实现，其大致逻辑和用户管理实现的逻辑基本一致，唯一的区别是功能有所增加。

9.2.4 角色管理

角色是实现权限控制的重要对象，可以理解为资源的集合，其和用户息息相关，一个用户可以有多个角色，一个角色也可以被多个用户拥有，所以用户和角色之间属于多对多的关系。

1．编写角色实体类

在 entity 包中创建 Role 实体对象，并添加用户实体和角色之间的关联，代码如下。

```
@Entity
@Table(name = "sys_roles")
public class Role extends BaseEntity {

    @Id
    @GenericGenerator(strategy = "uuid",name = "uuid")
    @GeneratedValue(generator="uuid")
    private String roleId;

    @Column(unique = true)
    private String roleName;
    @Column(unique = true)
    private String enName;    //英文名称
    private String roleDesc;
    private String roleValue;
}
```

修改用户实体类，添加其和角色的多对多关系，代码如下。

```
@Entity
@Table(name = "sys_users")
public class User extends BaseEntity {
    //….
    @ManyToMany(targetEntity = Role.class)
    @JoinColumn(name = "roleId")
    private Set<Role> roles;   //用户实体和角色的多对多关系
}
```

2．编写存储接口

在 repository 包中创建 RoleRepository 接口，用于处理角色信息的持久化。该接口比较简单，只需要继承 JpaRepository，并指明操作的实体类和 ID 所对应的数据类型即可。具体代码如下。

```
public interface RoleRepository extends JpaRepository<Role,String> {

}
```

3. 编写服务接口

在 service 包中添加 RoleService，其继承自 BaseService，并添加其他服务方法声明，代码如下。关于该接口中声明的方法的含义，可查看相应的注释。

```
public interface RoleService extends BaseService<Role,String> {

    /**
     * 获取该角色拥有的权限
     * @param id
     * @return
     */
    List<String> findRoleMenus(String id);

    /**
     * 解除角色和用户的关系
     * @param userId
     * @param roleId
     */
    void removeUser(String userId, String roleId);

    /**
     * 添加用户和角色的关系
     * @param roleId
     * @param userIds
     */
    void assignRole(String roleId, String[] userIds);
}
```

在 impl 包中定义该接口的实现类，并注入 Repository 层的代码，代码如下。

```
@Transactional(readOnly = true)
@Service("roleService")
public class RoleServiceImpl extends BaseServiceImpl<Role,String> implements RoleService{

    @Resource(name = "roleRepository")
    @Override
    public void setJpaRepository(JpaRepository<Role, String> jpaRepository) {
        super.setJpaRepository(jpaRepository);
    }

    @Transactional
    @Override
    public void delete(String s) {
        //1.清除角色和用户的关系
        String sql="DELETE from sys_users_roles where roles_role_id=?";
        //2.删除数据
        this.jdbcTemplate.update(sql,s);
        super.delete(s);
    }

    @Autowired
    private JdbcTemplate jdbcTemplate;
```

```java
    @Override
    public List<String> findRoleMenus(String id) {
        String sql="SELECT menus_id from sys_roles_menus where role_role_id=?";
        return this.jdbcTemplate.query(sql,new Object[]{id}
                ,new SingleColumnRowMapper<String>());
    }

    @Transactional(readOnly = false)
    @Override
    public void removeUser(String userId, String roleId) {
        String sql="DELETE from sys_users_roles where user_user_id=? and roles_role_id=?";
        this.jdbcTemplate.update(sql,userId,roleId);
    }

    @Transactional(readOnly = false)
    @Override
    public void assignRole(String roleId, String[] userIds) {
        //1.删除该角色的所有用户
        String sql="DELETE from sys_users_roles where roles_role_id=?";
        this.jdbcTemplate.update(sql,roleId);
        //2.添加关系
        sql="INSERT INTO sys_users_roles(user_user_id,roles_role_id) VALUES (?,?)";
        this.jdbcTemplate.batchUpdate(sql, new BatchPreparedStatementSetter() {
            @Override
            public void setValues(
                    PreparedStatement preparedStatement, int i) throws SQLException {
                preparedStatement.setString(1,userIds[i]);
                preparedStatement.setString(2,roleId);
            }

            @Override
            public int getBatchSize() {
                return userIds.length;
            }
        });
    }
}
```

对于删除角色而言，先要处理角色和用户的关联关系，将与该角色相关联的用户信息的关系断开，即先删除 sys_users_roles 表中存储的该角色相关联信息，再删除该角色信息，否则将出现外键约束异常。

4. 编写处理器

在 controller 包中编写 RoleController 类，以处理角色管理相关的请求，代码如下。

```java
@Controller
@RequestMapping("/sys/role")
public class RoleController {

    @Autowired
    private RoleService roleService;

    @RequestMapping(value = "/index",method = RequestMethod.GET)
    public String index(Model model, Pageable pageable){
        model.addAttribute("roles",this.roleService.findAll(pageable));
```

```java
        return "sys/role/roleList";
    }

    @RequestMapping(value = "/checkName")
    public @ResponseBody Boolean checkName(Role role){
        Example<Role> roleExample=Example.of(role);
        return !this.roleService.exists(roleExample);
    }

    @RequestMapping(value = "/checkEnName")
    public @ResponseBody Boolean checkEnName(Role role){
        Example<Role> roleExample=Example.of(role);
        return !this.roleService.exists(roleExample);
    }

    @Autowired
    private MenuService menuService;

    @RequestMapping(value = "/form",method = RequestMethod.GET)
    public String form(Model model,String id){
        Role role=null;
        if(StringUtils.isEmpty(id)){  //新增
            role=new Role();
        }else{   //修改
            role=this.roleService.getOne(id);
            model.addAttribute("menuIds",this.roleService.findRoleMenus(id));
        }
        model.addAttribute("menus",this.menuService.findAll());
        model.addAttribute("role",role);
        return "sys/role/roleForm";
    }

    @RequestMapping(value = "/save",method = RequestMethod.POST)
    public String save(Role role,String[] menuIds){
        if(menuIds!=null&&menuIds.length>0){
            Set<Menu> menus=new HashSet<>();
            Menu menu;
            for(String id:menuIds){
                menu=new Menu();
                menu.setId(id);
                menus.add(menu);
            }

            role.setMenus(menus);
        }
        role=this.roleService.saveOrUpdate(role);
        return "redirect:form?id="+role.getRoleId();
    }

    @RequestMapping(value = "/delete/{id}",method = RequestMethod.GET)
    public String delete(@PathVariable String id){
        this.roleService.delete(id);
        return "redirect:../index";
    }
```

```java
@Autowired
private UserService userService;

//权限分配
@RequestMapping("/assign")
public String assign(Role role,Model model){
    role = this.roleService.getOne(role.getRoleId());
    model.addAttribute("role",role);
    model.addAttribute("users",this.userService.findUsersByRole(role));
    return "sys/role/roleAssign";
}

@RequestMapping(value = "/outrole")
public String outrole(String userId,String roleId){
    this.roleService.removeUser(userId,roleId);
    return "redirect:assign?roleId="+roleId;
}

@RequestMapping(value = "/usertorole")
public String usertorole(Role role, Model model){
    model.addAttribute("role",this.roleService.getOne(role.getRoleId()));
    model.addAttribute("inUsers",this.userService.findUsersByRole(role));
    model.addAttribute("users",this.userService.findAll());
    return "sys/role/selectUserToRole";
}

@RequestMapping(value = "/assignrole")
public String assignRole(String roleId,String[] userIds){
    this.roleService.assignRole(roleId,userIds);
    return "redirect:assign?roleId="+roleId;
}

}
```

5. 编写列表界面

在 templates 文件夹中创建 sys/role 文件夹，并创建 roleList.ftl 文件，代码如下。

```html
<html>
<head>
    <title>角色管理</title>
    <meta name="decorator" content="default"/>
    <#include "../../layout/header.ftl">
    <style>
        #contentTable{
            font-size: 13px;
        }
    </style>
</head>
<body>
<ul class="nav nav-tabs">
    <li class="active"><a href="${ctx}/sys/role/index">角色列表</a></li>
    <@shiro.hasPermission name="sys:role:form">
       <li><a href="${ctx}/sys/role/form">角色添加</a></li>
    </@shiro.hasPermission>
</ul>
<table id="contentTable" class="table table-striped table-bordered table-condensed">
```

```
            <tr>
<th>角色名称</th>
<th>英文名称</th>
<th>角色描述</th>
<th>角色值</th>
<@shiro.hasPermission name="sys:role:edit">
<th>操作</th>
</@shiro.hasPermission>
</tr>
        <#list roles.content as role>
            <tr>
                <td><a href="${ctx}/sys/role/form?id=${role.roleId}">${role.roleName!}</a> </td>
                <td><a href="${ctx}/sys/role/form?id=${role.roleId}">${role.enName!}</a></td>
                <td>${role.roleDesc!}</td>
                <td>${role.roleValue!}</td>
                <@shiro.hasPermission name="sys:role:edit">
                    <td>
                        <a href="${ctx}/sys/role/assign?roleId=${role.roleId}">权限分配</a>
                        <a href="${ctx}/sys/role/form?id=${role.roleId}">修改</a>
                        <a href="${ctx}/sys/role/delete/${role.roleId}"
                      onclick="return confirmx('确认要删除该角色吗？', this.href)">
                            删除</a>
                    </td>
                </@shiro.hasPermission>
            </tr>
    </#list>
</table>
</body>
</html>
```

6. 编写添加、修改界面

在 sys/role 文件夹中创建 roleForm.ftl 文件，其主要通过 RoleController 中的 form 方法实现业务处理，代码如下。

```
<html>
<head>
    <title>角色管理</title>
    <meta name="decorator" content="default"/>
    <#include "../../layout/header.ftl">
</head>
<body>
<ul class="nav nav-tabs">
    <li><a href="${ctx}/sys/role/index">角色列表</a></li>
    <li class="active">
       <a href="${ctx}/sys/role/form?id=${role.roleId!}">角色
       <@shiro.hasPermission name="sys:role:edit">
       ${(role.roleId??)?then('修改','添加')}
       @shiro.hasPermission
       <@shiro.lacksPermission name="sys:role:edit">查看</@shiro.lacksPermission></a></li>
</ul>
<br/>
<form id="inputForm" action="${ctx}/sys/role/save" method="post" class="form-horizontal">
    <input name="roleId" type="hidden" value="${role.roleId!}"/>
```

```html
<div class="control-group">
    <label class="control-label">角色名称:</label>
    <div class="controls">
        <input name="roleName" htmlEscape="false" maxlength="50"
                        class="required" value="${role.roleName!}"/>
        <span class="help-inline"><font color="red">*</font> </span>
    </div>
</div>
<div class="control-group">
    <label class="control-label">英文名称:</label>
    <div class="controls">
        <input name="enName" htmlEscape="false"
 maxlength="50" class="required" value="${role.enName!}"/>
        <span class="help-inline"><font color="red">*</font> </span>
    </div>
</div>
<div class="control-group">
    <label class="control-label">角色值:</label>
    <div class="controls">
        <input name="roleValue" htmlEscape="false" maxlength="50"
                        class="required" value="${role.roleValue!}"/>
        <span class="help-inline"><font color="red">*</font> </span>
    </div>
</div>
<div class="control-group">
    <label class="control-label">角色授权:</label>
    <div class="controls">
        <div id="menuTree" class="ztree" style="margin-top:3px;float:left;"></div>
        <input type="hidden" name="menuIds" id="menuIds">
    </div>
</div>
<#if message??>
    <div id="messageBox" class="alert alert-error">
        <button data-dismiss="alert" class="close">×</button>
    ${message}
    </div>
</#if>
<#-- 关闭错误提示 -->
<script type="text/javascript">top.$.jBox.closeTip();</script>
<div class="control-group">
    <label class="control-label">角色描述:</label>
    <div class="controls">
        <textarea name="roleDesc" htmlEscape="false" rows="3"
                        maxlength="200" class="input-xlarge">
                        ${role.roleDesc!}</textarea>
    </div>
</div>
<div class="form-actions">
    <@shiro.hasPermission name="sys:role:edit">
        <input id="btnSubmit" class="btn btn-primary"
  type="submit" value="保 存"/> 
    </@shiro.hasPermission>
    <input id="btnCancel" class="btn"
  type="button" value="返 回" onclick="history.go(-1)"/>
```

```
        </div>
</form>
</body>
<link href="${ctx}/jquery-ztree/3.5.12/css/zTreeStyle/zTreeStyle.min.css"
      rel="stylesheet" type="text/css"/>
<script
        src="${ctx}/jquery-ztree/3.5.12/js/jquery.ztree.all-3.5.min.js"
type="text/javascript"></script>
<script type="text/javascript">
    $(document).ready(function () {
        $("#roleName").focus();
        $("#inputForm").validate({
            rules: {
                roleValue: {required: true},
            <#if !role.roleId??>
                roleName: {
                    remote: {
                        url: "${ctx}/sys/role/checkName",
                        type: 'POST',
                        data: {
                            roleName: function () {
                                return $("#inputForm input[name='roleName']").val()
                            }
                        }
                    }
                },
                enName: {
                    remote: {
                        url: "${ctx}/sys/role/checkEnName",
                        type:'POST',
                        data: {
                            enName: function () {
                                return $("#inputForm input[name='enName']").val()
                            }
                        }
                    }
                }
            </#if>
            },
            messages: {
                roleName: {remote: "角色名已存在"},
                enName: {remote: "角色名已存在"},
                roleValue: {required: "角色值必填"}
            },
            submitHandler: function (form) {
                var ids = [], nodes = tree.getCheckedNodes(true);
                for(var i=0; i<nodes.length; i++) {
                    ids.push(nodes[i].id);
                }
                $("#menuIds").val(ids);
                loading('正在提交，请稍等...');
                form.submit();
            },
            errorContainer: "#messageBox",
            errorPlacement: function (error, element) {
```

```
                $("#messageBox").text("输入有误,请先更正。");
                if (element.is(":checkbox")
                    || element.is(":radio")
                    || element.parent().is(".input-append")) {
                        error.appendTo(element.parent().parent());
                } else {
                        error.insertAfter(element);
                }
            }
        });

        //初始化权限
        var setting = {
            check: {enable: true, nocheckInherit: true}, view: {selectedMulti: false},
            data: {simpleData: {enable: true}},
            callback: {
                beforeClick: function (id, node) {
                    tree.checkNode(node, !node.checked, true, true);
                    return false;
                }
            }
        };

        // 用户-菜单
        var zNodes=[
            <#list menus as menu>
                {id:"${menu.id}",
                 pId:"${(menu.parent??)?then(menu.parent.id,'0')}",
                 name:"${(menu.parent??)?then(menu.title,'权限列表')}"},
            </#list>
        ];
        //初始化树结构
        var tree = $.fn.zTree.init($("#menuTree"), setting, zNodes);
        //不选择父节点
        tree.setting.check.chkboxType = { "Y" : "ps", "N" : "s" };
        //默认选择节点
        <#if menuIds??>
            <#list menuIds as id>
                var node = tree.getNodeByParam("id", '${id}');
                try{tree.checkNode(node, true, false);}catch(e){}
            </#list>
        </#if>
        //默认展开全部节点
        tree.expandAll(true);
    });
</script>
</html>
```

7. 编写用户授权界面

在 sys/role 文件夹中创建 roleAssign.ftl 文件,用于展示用户和角色之间的关联信息,主要通过 RoleController 中的 assign 方法实现页面展示。具体代码如下。

```
<html>
<head>
    <title>分配角色</title>
    <meta name="decorator" content="default"/>
    <#include "../../layout/header.ftl">
```

```html
    <style>
        #contentTable{
            font-size: 13px;
        }
    </style>
</head>
<body>
<ul class="nav nav-tabs">
    <li><a href="${ctx}/sys/role/index">角色列表</a></li>
    <li class="active">
        <a href="${ctx}/sys/role/assign?id=${role.roleId!}">
        角色分配</a></li>
</ul>
<div class="container-fluid breadcrumb">
    <div class="row-fluid span12">
        <span class="span4">角色名称：<b>${role.roleName!}</b></span>
        <span class="span4">英文名称：${role.enName!}</span>
    </div>
</div>
<div class="breadcrumb">
    <#-- 分配用户信息 -->
    <form    id="assignRoleForm"    action="${ctx}/sys/role/assignrole"    method="post"
class="hide">
        <input type="hidden" name="roleId" value="${role.roleId}"/>
        <input id="idsArr" type="hidden" name="userIds" value=""/>
    </form>
    <input id="assignButton" class="btn btn-primary" type="submit" value="分配角色"/>
    <script type="text/javascript">
        $("#assignButton").click(function(){
            top.$.jBox.open(
                "iframe:${ctx}/sys/role/usertorole?roleId=${role.roleId}",
                "分配角色",
                810,
                $(top.document).height()-240,
                {
                    buttons:{
                        "确定分配":"ok",
                        "清除已选":"clear",
                        "关闭":true},
                    bottomText:"通过选择部门，为列出的人员分配角色。",
                    submit:
                      function(v, h, f){
                            var pre_ids = h.find("iframe")[0].contentWindow.
pre_ids;
                            var ids = h.find("iframe")[0].contentWindow.ids;
                            //nodes = selectedTree.getSelectedNodes();
                            if (v=="ok"){
                                //删除''的元素
                            if(ids[0]==''){
                                ids.shift();
                                pre_ids.shift();
                            }
                            if(pre_ids.sort().toString() == ids.sort().toString()){
                                top.$.jBox.tip(
                                    "未给角色【${role.roleName}】分配新成员！"
```

```
                                    , 'info');
                                return false;
                            };
                            //执行保存操作
                            loading('正在提交，请稍等...');
                            var idsArr = "";
                            for (var i = 0; i<ids.length; i++) {
                                idsArr = (idsArr + ids[i]) + (((i + 1) == ids.length) ?
'':',');
                            }
                            $('#idsArr').val(idsArr);
                            $('#assignRoleForm').submit();
                            return true;
                        } else if (v=="clear"){
                            h.find("iframe")[0].contentWindow.clearAssign();
                            return false;
                        }
                    }, loaded:function(h){
                        $(".jbox-content", top.document).css("overflow-y","hidden");
                    }
                });
            });
        </script>
    </div>
<script type="text/javascript">top.$.jBox.closeTip();</script>
<table id="contentTable" class="table table-striped table-bordered table-condensed">
    <thead>
        <tr>
            <th>登录名</th>
            <th>姓名</th>
            <th>电话</th>
            <th>邮件</th>
            <th>操作</th>
        </tr>
    </thead>
    <tbody>
    <#list users as user>
        <tr>
            <td><a href="${ctx}/sys/user/form?id=${user.userId}">${user.userName}
</a></td>
            <td>${user.realName!}</td>
            <td>${user.phoneNum!}</td>
            <td>${user.email!}</td>
            <td>
                <a href="${ctx}/sys/role/outrole?userId=${user.userId}&roleId=$
{role.roleId}"
                    onclick="return confirmx('确认要将用户
                        <b>[${user.userName}]</b>从<b>[${role.roleName}]</b>
                        角色中移除吗？', this.href)">移除</a>
            </td>
        </tr>
    </#list>
    </tbody>
</table>
</body>
```

</html>

角色信息的分配是以弹窗的方式进行的。单击分配角色按钮将显示弹窗，并对该角色将要分配给的用户的信息进行展示，该请求主要由 RoleController 中的 assignrole 方法进行处理。这里需要编写弹窗显示界面，在 sys/role 文件夹中编写 selectUserToRole.ftl 文件，代码如下。

```html
<html>
<head>
    <title>分配角色</title>
    <meta name="decorator" content="blank"/>
    <#include "../../layout/header.ftl">
</head>
<body>
<div id="assignRole" class="row-fluid span12">
    <div class="span3">
        <p>待选人员：</p>
        <div id="userTree" class="ztree"></div>
    </div>
    <div class="span3" style="padding-left:16px;border-left: 1px solid #A8A8A8;">
        <p>已选人员：</p>
        <div id="selectedTree" class="ztree"></div>
    </div>
</div>
<link href="${ctx}/jquery-ztree/3.5.12/css/zTreeStyle/zTreeStyle.min.css"
                                                    rel="stylesheet" type="text/css"/>
<script src="${ctx}/jquery-ztree/3.5.12/js/jquery.ztree.all-3.5.min.js"
type="text/javascript"></script>
<script type="text/javascript">
    var selectedTree;   //zTree的已选择对象
    // 进行初始化
    $(document).ready(function(){
        selectedTree = $.fn.zTree.init($("#selectedTree"), setting, selectedNodes);
        $.fn.zTree.init($("#userTree"), setting, allUsers);
    });
    var setting = {
                view: {
                    selectedMulti:false,
                    nameIsHTML:true,
                    showTitle:false,
                    dblClickExpand:false
                    },
                data: {simpleData: {enable: true}},
                callback: {onClick: treeOnClick}};

    //所有用户信息
    var allUsers=[
    <#list users as user>
        {id:"${user.userId}",
            pId:"0",
            name:"${user.userName}"
        },
    </#list>
    ]

    //缓存原始已选择的用户数据
```

```
var pre_selectedNodes =[
<#list inUsers as user>
    {id:"${user.userId}",
        pId:"0",
        name:"<font color='red' style='font-weight:bold;'>${user.userName}</font>"
    },
</#list>
];

var selectedNodes =[
<#list inUsers as user>
    {id:"${user.userId}",
        pId:"0",
        name:"<font color='red' style='font-weight:bold;'>${user.userName}</font>"
    },
</#list>
];

//保存原始数据
var pre_ids = [
<#list inUsers as user>
    '${user.userId}',
</#list>
];
var ids = [
<#list inUsers as user>
    '${user.userId}',
</#list>
];

//选择选项并回调
function treeOnClick(event, treeId, treeNode, clickFlag){
    $.fn.zTree.getZTreeObj(treeId).expandNode(treeNode);
    if("userTree"==treeId){
        if($.inArray(String(treeNode.id), ids)<0){
            selectedTree.addNodes(null, treeNode);
            ids.push(String(treeNode.id));
        }
    };
    if("selectedTree"==treeId){
        if($.inArray(String(treeNode.id), pre_ids)<0){
            selectedTree.removeNode(treeNode);
            ids.splice($.inArray(String(treeNode.id), ids), 1);
        }else{
            top.$.jBox.tip("角色原有成员不能清除！", 'info');
        }
    }
};
function clearAssign(){
    var submit = function (v, h, f) {
        if (v == 'ok'){
            var tips="";
            if(pre_ids.sort().toString() == ids.sort().toString()){
                tips = "未给角色【${role.roleName}】分配新成员！";
            }else{
```

```
                            tips = "已选人员清除成功！";
                    }
                    ids=pre_ids.slice(0);
                    selectedNodes=pre_selectedNodes;
                    $.fn.zTree.init($("#selectedTree"), setting, selectedNodes);
                    top.$.jBox.tip(tips, 'info');
                } else if (v == 'cancel'){
                    //取消清除操作
                    top.$.jBox.tip("取消清除操作！", 'info');
                }
                return true;
            };
            tips="确定清除角色【${role.roleName}】下的已选人员？";
            top.$.jBox.confirm(tips, "清除确认", submit);
        };
</script>
</body>
</html>
```

通过弹窗展示用户角色分配处理，并监听窗口的单击事件，将赋予的角色 ID 和用户 ID 数组提交给 RoleController 中的 assignRole 方法，以实现用户角色的分配。

9.2.5 菜单管理

实现用户、角色信息的管理后，要对权限的细粒度进行控制，需要对系统中所有的资源进行分类。最常见的处理方式是每个资源对应一个菜单地址（有的资源实质上是按钮），下面将讲解菜单信息的维护及其和角色的绑定。

1. 编写菜单实体类

在 entity 包中创建 Menu 实体对象，因为菜单存在多级问题，即一个父菜单拥有多个子菜单，故菜单与其自身也是一对多的关系。具体代码如下。

```
@Entity
@Table(name = "sys_menus")
public class Menu extends BaseEntity {

    @Id
    @GenericGenerator(strategy = "uuid",name = "uuid")
    @GeneratedValue(generator="uuid")
    private String id;
    private String title;//名称

    @Column(unique = true)
    private String url;//地址
    private Integer sort;//排序
    private String icon;//图标
    private String permission;//Shiro权限
    private String remarks;//说明
    private Date createDate;//创建时间
    private Boolean isShow;//是否可见
    private String createBy;//创建人
    private String target;//目标

    @ManyToOne(targetEntity = Menu.class)
    @JoinColumn(name = "parentId")
```

```java
    private Menu parent;//父菜单
}
```

编写完菜单实体之后修改角色实体，关联其和菜单的多对多关系，代码如下。

```java
@Entity
@Table(name = "sys_roles")
public class Role extends BaseEntity {

    @ManyToMany(targetEntity = Menu.class,cascade = CascadeType.PERSIST)
    @JoinColumn(name = "id")
    private Set<Menu> menus;//角色实体和菜单的多对多关系
}
```

2. 编写存储接口

在 repository 包中创建 MenuRepository 接口，代码如下。

```java
public interface MenuRepository  extends JpaRepository<Menu,String>{

    /**
     * 查询父菜单为空的菜单
     * @return
     */
    List<Menu> findMenuByParentNotNullOrderBySortAsc();

    /**
     * 查询子菜单
     * @param parentId
     * @return
     */
    List<Menu> findMenuByParentIdOrderBySortAsc(String parentId);

    /**
     * 查询该父菜单中排序的最大值
     * @param id
     * @return
     */
    @Query("select max(sort) from Menu where parent.id=?1")
    Integer findMaxSortById(String id);
}
```

3. 编写服务接口

在 service 包中创建 MenuService 接口，其继承自 BaseService，并声明其服务方法，详细信息可查看注释。具体代码如下。

```java
public interface MenuService extends BaseService<Menu,String>{

    /**
     * 获取树形菜单
     * @return
     */
    List<Map<String,Object>> findTreeSelectMenus();

    /**
     * 获取treeTable的菜单数据
     * @return
     */
    List<Menu> findTreeTableMenus();

    /**
```

```
 * 通过父菜单的ID获取其子节点
 * @return
 */
List<Menu> findMenusByParentId(String parentId);

/**
 * 获取左侧导航菜单的数据
 * @param parentId
 * @return
 */
List<TreeMenu> findLeftMenus(String parentId);

/**
 * 查询最大的排序值
 * @param parentId
 * @return
 */
Integer findMaxSortById(String parentId);
}
```

在 impl 包中编写该接口的实现类,此类与 UserServiceImpl 比较类似,除了注入 Repository 接口之外还注入了 JdbcTemplate 对象,因为在某些情况下直接使用 JDBC 更为方便。具体代码如下。

```
@Transactional(readOnly = true)
@Service("menuService")
public class MenuServiceImpl extends BaseServiceImpl<Menu,String> implements MenuService{

    @Autowired
    private MenuRepository menuRepository;

    @Resource(name = "menuRepository")
    @Override
    public void setJpaRepository(JpaRepository<Menu, String> jpaRepository) {
        super.setJpaRepository(jpaRepository);
    }

    @Transactional(readOnly = false)
    @Override
    public void delete(String s) {
        //1.清除菜单和角色的关系
        this.deleteRoleByMenuId(s);
        //2.删除子节点
        deleteChildren(s);
        //3.删除自己
        if(this.menuRepository.exists(s)){
            this.menuRepository.delete(s);
        }
    }

    //删除子菜单
    private void deleteChildren(String parentId){
        List<Menu> menus =
                this.menuRepository.findMenuByParentIdOrderBySortAsc(parentId);
        if(menus!=null&&menus.size()>0){   //判断是否为子菜单
            for(Menu menu:menus){
                delete(menu.getId());    //递归删除子节点
                if(this.menuRepository.exists(menu.getId())) {
```

```java
                    this.deleteRoleByMenuId(parentId);
                    super.delete(menu.getId());   //删除自己
                }
            }
        }else{
            this.deleteRoleByMenuId(parentId);
            super.delete(parentId);
        }
    }

    //处理子菜单的角色问题
    private void deleteRoleByMenuId(String menuId){
        String sql="delete from sys_roles_menus where menus_id=?";
        this.jdbcTemplate.update(sql,menuId);
    }

    @Override
    public List<Menu> findMenusByParentId(String parentId) {
        return this.menuRepository.findMenuByParentIdOrderBySortAsc(parentId);
    }

    @Override
    public List<TreeMenu> findLeftMenus(String parentId) {
        //1.获取二级菜单
        List<Menu> menus =
                this.menuRepository.findMenuByParentIdOrderBySortAsc(parentId);
        List<TreeMenu> result=new ArrayList<>();
        TreeMenu treeMenu=null;
        for(Menu menu:menus){
            treeMenu=new TreeMenu();
            //2.复制数据
            BeanUtils.copyProperties(menu,treeMenu);
            //3.设置子节点
            treeMenu.setChildren(
                    this.menuRepository.findMenuByParentIdOrderBySortAsc(menu.getId()));
            result.add(treeMenu);
            treeMenu=null;
        }
        return result;
    }

    @Autowired
    private JdbcTemplate jdbcTemplate;

    @Override
    public List<Map<String, Object>> findTreeSelectMenus() {
        String sql =
                "SELECT sm.id,sm.title as 'name',sm.parent_id as 'pId' from sys_menus sm"
                +"where sm.is_show=true order by sort";
        return this.jdbcTemplate.query(sql,new ColumnMapRowMapper());
    }

    @Override
    public Integer findMaxSortById(String id) {
```

```java
            return this.menuRepository.findMaxSortById(id);
    }

    @Override
    public List<Menu> findTreeTableMenus() {
        List<Menu> result=new ArrayList<>();
        List<Menu> source = this.menuRepository.findMenuByParentNotNullOrderBySortAsc();
        sortList(result,source,Menu.getRootMenu(),true);
        return result;
    }

    /**
     * 将属于一组的菜单放到一起
     * @param result
     * @param source
     * @param parentId
     * @param cascade
     */
    public void sortList(
                    List<Menu> result, List<Menu> source, String parentId, boolean cascade) {
        for(int i=0;i<source.size();i++){
            Menu e = source.get(i);
            //查找所有的顶级菜单
            if(e.getParent().getId().equals(parentId)){
                result.add(e);
                if(cascade){
                    // 判断是否还有子节点, 有则继续获取子节点
                    for (int j=0; j<source.size(); j++){
                        Menu child = source.get(j);
                        if (child.getParent().getId().equals(e.getId())){
                            sortList(result, source, e.getId(), true);
                            break;
                        }
                    }
                }
            }
        }
    }
}
```

4. 编写处理器

在 controller 包中创建 MenuController，以处理和菜单相关的请求，代码如下。

```java
@Controller
@RequestMapping("/sys/menu")
public class MenuController {

    @Autowired
    private MenuService menuService;

    @RequiresPermissions("sys:menu:index")
    @RequestMapping(value = "/index",method = RequestMethod.GET)
    public String index(Model model){
        List<Menu> menus = this.menuService.findTreeTableMenus();
        model.addAttribute("menus",menus);
        return "sys/menu/menuList";
```

```java
    }

    @RequestMapping(value = "/form",method = RequestMethod.GET)
    public String form(Model model,Menu menu){
        if(StringUtils.isEmpty(menu.getId())){   //新增或添加子菜单
            if(menu.getParent()==null) {   //新增子菜单
                menu.setParent(this.menuService.getOne(Menu.getRootMenu()));
            }else{
                menu.setParent(this.menuService.getOne(menu.getParent().getId()));
            }
            //初始化该菜单的排序值为当前父菜单的最大值
            menu.setSort(menuService.findMaxSortById(menu.getParent().getId())+1);
        }else{
            menu=this.menuService.getOne(menu.getId());
        }
        model.addAttribute("menu",menu);
        return "sys/menu/menuForm";
    }

    @RequestMapping(value = "/save",method = RequestMethod.POST)
    public String save(Menu menu,Model model){
        User user=(User)SecurityUtils.getSubject().getPrincipal();
        menu.setCreateBy(user.getUserId());
        menu=this.menuService.saveOrUpdate(menu);
        model.addAttribute("menu",menu);
        return  "sys/menu/menuForm";
    }

    @RequestMapping(value = "/delete",method = RequestMethod.GET)
    public String delete(String id){
        this.menuService.delete(id);
        return "redirect:index";
    }

    @RequestMapping("/treeData")
    public @ResponseBody List<Map<String,Object>> treeData(){
        return this.menuService.findTreeSelectMenus();
    }

    @RequestMapping(value = "/tree")
    public String tree(String parentId,Model model){
        model.addAttribute("parentId",parentId);
        model.addAttribute("leftMenus",this.menuService.findLeftMenus(parentId));
        return "sys/menu/menuTree";
    }
}
```

5. 编写列表界面

在 templates 文件夹中创建 sys/menu 文件夹，并在 sys/menu 中创建 menuList.ftl 文件，其主要通过 MenuController 的 index 方法实现数据展示。具体代码如下。

```
<html>
<head>
    <title>菜单管理</title>
    <meta name="decorator" content="default"/>
    <#include "../../layout/header.ftl">
```

```html
    <style>
        #treeTable{
            font-size: 13px;
        }
    </style>
    <link href="${ctx}/treeTable/themes/vsStyle/treeTable.min.css"
                                            rel="stylesheet" type="text/css">
    <script    src="${ctx}/treeTable/jquery.treeTable.min.js"    type="text/javascript">
</script>
    <script type="text/javascript">
        $(document).ready(function () {
            $("#treeTable").treeTable({expandLevel : 3}).show();
        });
    </script>
</head>
<body>
<ul class="nav nav-tabs">
    <li class="active"><a href="${ctx}/sys/menu/index">菜单列表</a></li>
    <@shiro.hasPermission name="sys:menu:form">
        <li><a href="${ctx}/sys/menu/form">菜单添加</a></li>
    </@shiro.hasPermission>
</ul>
<form id="listForm" method="post">
    <table id="treeTable" class="table table-striped table-bordered table-condensed">
        <thead>
                <tr>
                            <th>名称</th>
                            <th>链接</th>
                            <th style="text-align:center;">排序</th>
                            <th>可见</th>
                            <th>权限标识</th>
                            <@shiro.hasPermission name="sys:menu:edit">
                            <th>操作</th>
                            </@shiro.hasPermission></tr></thead>
        <tbody>
        <#list menus as menu>
            <tr id="${menu.id}" pId="${(menu.parent??)?then(menu.parent.id!,'0')}">
                <td nowrap>
                            <i class="icon-${(menu.icon??)?then(menu.icon,'hide')}"></i>
                        <ahref="${ctx}/sys/menu/form?id=${menu.id!}">
                                ${menu.title!}</a>
                </td>
                <td title="${menu.url!}">${menu.url!}</td>
                <td style="text-align:center;">
                    ${menu.sort!}
                </td>
                <td>${menu.isShow ?then('显示','隐藏')}</td>
                <td title="${menu.permission!}">${menu.permission!}</td>
                <@shiro.hasPermission name="sys:menu:edit">
                    <td nowrap>
                        <a href="${ctx}/sys/menu/form?id=${menu.id!}">修改</a>
                        <a href="${ctx}/sys/menu/delete?id=${menu.id!}"
                                onclick=
                        "return confirmx('要删除该菜单及所有子菜单项吗？', this.href)">
                                删除</a>
```

```
                    <a href="${ctx}/sys/menu/form?parent.id=${menu.id!}">
                        添加下级菜单</a>
                </td>
            </@shiro.hasPermission>
        </tr>
    </#list>
    </tbody>
  </table>
</form>
</body>
</html>
```

6. 编写菜单添加界面

在相应的文件夹中创建 menuForm.ftl 文件，用于处理菜单的添加和修改操作，具体代码如下。由于菜单和其自己属于一对多的关系，因此需要选择父菜单，这里也以弹窗的方式进行处理。

```
<html>
<head>
    <title>菜单管理</title>
    <meta name="decorator" content="default"/>
    <#include "../../layout/header.ftl">
    <#include "../../commons/tags.ftl">
    <script type="text/javascript">
        $(document).ready(function() {
            $("#name").focus();
            $("#inputForm").validate({
                submitHandler: function(form){
                    loading('正在提交，请稍等...');
                    form.submit();
                },
                errorContainer: "#messageBox",
                errorPlacement: function(error, element) {
                    $("#messageBox").text("输入有误，请先更正。");
                    if   (element.is(":checkbox")||element.is(":radio")||element.parent().is(".input-append")){
                        error.appendTo(element.parent().parent());
                    } else {
                        error.insertAfter(element);
                    }
                }
            });
        });
    </script>
</head>
<body>
<ul class="nav nav-tabs">
    <li><a href="${ctx}/sys/menu/index">菜单列表</a></li>
    <li class="active">
<a href="${ctx}/sys/menu/form">菜单
<@shiro.hasPermission name="sys:menu:edit">
${(menu.id??)?then('修改','添加')}
</@shiro.hasPermission>
<@shiro.lacksPermission name="sys:menu:edit">查看</@shiro.lacksPermission></a></li>
</ul><br/>
<form id="inputForm" action="${ctx}/sys/menu/save" method="post" class="form-horizontal">
```

```html
<input type="hidden" name="id" value="${menu.id!}">
<div class="control-group">
    <label class="control-label">上级菜单:</label>
    <div class="controls">
        <@treeselect id="menu"
                    name="parent.id" value="${(menu.parent??)?then(menu.parent.id,'')}"
                    labelName="parent.title"
                    labelValue="${(menu.parent??)?then(menu.parent.title,'')}"
                    title="菜单"
                    url="/sys/menu/treeData"
                    extId="${menu.id!}"
                    cssStyle="height: 30px;background-color: #ffffff"/>
    </div>
</div>
<script type="text/javascript">top.$.jBox.closeTip();</script>
<div class="control-group">
    <label class="control-label">名称:</label>
    <div class="controls">
        <input name="title" htmlEscape="false" maxlength="50"
                        class="required input-xlarge" value="${menu.title!}"/>
        <span class="help-inline"><font color="red">*</font> </span>
    </div>
</div>
<div class="control-group">
    <label class="control-label">链接:</label>
    <div class="controls">
        <input name="url" htmlEscape="false" maxlength="2000"
                            class="input-xxlarge" value="${menu.url!}"/>
        <span class="help-inline">单击菜单跳转的页面</span>
    </div>
</div>
<div class="control-group">
    <label class="control-label">目标:</label>
    <div class="controls">
        <input name="target" htmlEscape="false" maxlength="10"
                            class="input-small" value="${menu.target!}"/>
        <span class="help-inline">链接地址打开的目标窗口，默认：mainFrame</span>
    </div>
</div>
<div class="control-group">
    <label class="control-label">图标:</label>
    <div class="controls">
        <@iconselect id="icon" name="icon" value="${menu.icon!}"/>
    </div>
</div>
<div class="control-group">
    <label class="control-label">排序:</label>
    <div class="controls">
        <input name="sort" htmlEscape="false" maxlength="50"
                            class="required digits input-small"
                                        value="${menu.sort!}"/>
        <span class="help-inline">排列顺序，升序。</span>
    </div>
</div>
```

```html
        <div class="control-group">
            <label class="control-label">可见:</label>
            <div class="controls">
                <input name="isShow" type="radio" htmlEscape="false"
                        value="1" class="required"
                        ${(menu.isShow??)?then(menu.isShow?then('checked',''),'checked')}>
                        显示
                <input name="isShow" type="radio" htmlEscape="false"
                        value="0" class="required"
                        ${(menu.isShow??)?then((!menu.isShow)?then('checked',''),'')}>
                        隐藏
                <span class="help-inline">该菜单或操作是否显示到系统菜单中</span>
            </div>
        </div>
        <div class="control-group">
            <label class="control-label">权限标识:</label>
            <div class="controls">
                <input name="permission" htmlEscape="false" maxlength="100"
                        class="required input-xxlarge" value="${menu.permission!}"/>
                <span class="help-inline">
                        控制器中定义的权限标识，如：@RequiresPermissions("权限标识")
                </span>
            </div>
        </div>
        <div class="control-group">
            <label class="control-label">备注:</label>
            <div class="controls">
                <textarea name="remarks" htmlEscape="false" rows="3"
                        maxlength="200" class="input-xxlarge">${menu.remarks!}
                </textarea>
            </div>
        </div>
        <div class="form-actions">
            <@shiro.hasPermission name="sys:menu:edit">
                <input id="btnSubmit" class="btn btn-primary" type="submit"
                                                            value="保 存"/> 
            </@shiro.hasPermission>
            <input id="btnCancel" class="btn" type="button"
                                            value="返 回" onclick="history.go(-1)"/>
        </div>
</form>
</body>
</html>
```

细心的读者会发现，在使用的标签中存在"<@treeselect>"标签对，这里采用了 FreeMarker 中的自定义命令，定义了一个用于展现属性菜单的弹窗，在 templates 文件夹中创建 commons/tags 文件夹，并创建 tags.ftl 文件，用于存放定义的命令，代码如下。

```
<#-- 定义图标选择 -->
<#macro iconselect id name value>
<i id="${id}Icon" class="icon-${(value??)?then(value,' hide')}">
</i> <label id="${id}IconLabel">${(value??)?then(value,'无')}</label> 
<input id="${id}" name="${name}" type="hidden" value="${value}"/>
<a id="${id}Button" href="javascript:" class="btn">选择</a>  
<script type="text/javascript">
```

```
            $("#${id}Button").click(function(){
                top.$.jBox.open(
                    "iframe:${ctx}/sys/tag/iconselect?value="+$("#${id}").val(),
                    "选择图标",
                    700,
                    $(top.document).height()-180,
                    {
                        buttons:
                            {
                             "确定":"ok",
                             "清除":"clear",
                             "关闭":true
                             },
                        submit:function(v, h, f){
                          if (v=="ok"){
                                var icon = h.find("iframe")[0].contentWindow.$("#icon").val();
                                icon = $.trim(icon).substr(5);
                                $("#${id}Icon").attr("class", "icon-"+icon);
                                $("#${id}IconLabel").text(icon);
                                $("#${id}").val(icon);
                          }else if (v=="clear"){
                                $("#${id}Icon").attr("class", "icon- hide");
                                $("#${id}IconLabel").text("无");
                                $("#${id}").val("");
                          }
                        }, loaded:function(h){
                            $(".jbox-content", top.document).css("overflow-y","hidden");
                        }
                });
        });
</script>
</#macro>

<#-- 定义菜单选择，默认值的参数必须在参数列表的最后-->
<#macro treeselect id name value labelName labelValue url title
                cssClass=""
                extId=""
                cssStyle=""
                module=""
                checked=false
                isAll=true
                notAllowSelectRoot=false
                notAllowSelectParent=false
                selectScopeModule=false
                allowClear=false
                allowInput=false
                smallBtn=false
                hideBtn=false
                disabled=""
                dataMsgRequired=""
>
<div class="input-append">
    <input id="${id}Id" name="${name}" class="${cssClass}" type="hidden" value="${value}"/>
    <input id="${id}Name" name="${labelName}"
```

```
                    ${allowInput?then('','readonly="readonly"')}
                    type="text" value="${labelValue}"
                    data-msg-required="${dataMsgRequired}"
                    class="${cssClass}" style="${cssStyle}"/>
    <a id="${id}Button" href="javascript:"
                    class="btn ${disabled} ${hideBtn ? then('hide' , '')}"
                    style="${smallBtn?then('padding:4px 2px;','')}">
                     <i class="icon-search"></i> </a>  
</div>
<script type="text/javascript">
    $("#${id}Button, #${id}Name").click(function(){
        //是否限制选择，如果限制，则应设置为disabled
        if ($("#${id}Button").hasClass("disabled")){
            return true;
        }
        //正常打开
        top.$.jBox.open(
            "iframe:${ctx}/sys/tag/treeselect?url="+encodeURIComponent("${url}")
            +"&module=${module}&checked=${checked?c}&extId=${extId}&isAll=${isAll?c}",
            "选择${title}",
            300,
            420,
            {
                ajaxData:
                    {
                        selectIds: $("#${id}Id").val()},
                        buttons:{
                            "确定":"ok",
                            ${allowClear?then("\"清除\":\"clear\", ","")}"关闭":true
                        },
                        submit:function(v, h, f){
                            if (v=="ok"){
                                var tree =
                                    h.find("iframe")[0].contentWindow.tree;//h.find("iframe").contents();
                                var ids = [], names = [], nodes = [];
                                if ("${checked?c}" == "true"){
                                    nodes = tree.getCheckedNodes(true);
                                }else{
                                    nodes = tree.getSelectedNodes();
                                }
                                for(var i=0; i<nodes.length; i++) {
                                    <#if checked && notAllowSelectParent>
                                        if (nodes[i].isParent){
                                            continue; //如果为复选框选择，则过滤掉父节点
                                        }
                                    </#if>
                                    <#if notAllowSelectRoot>
                                        if (nodes[i].level == 0){
                                            top.$.jBox.tip(
                                                "不能选择根节点("+nodes[i].name+"),请重新选择。");
                                            return false;
                                        }
                                    </#if>
                                    <#if notAllowSelectParent>
                                        if (nodes[i].isParent){
                                            top.$.jBox.tip(
```

```
                                "不能选择父节点 ("+nodes[i].name+")，请重新选择。");
                                return false;
                            }
                        </#if>
                        <#if (module??) && selectScopeModule>
                            if (nodes[i].module == ""){
                                top.$.jBox.tip(
                                "不能选择公共模型 ("+nodes[i].name+")，请重新选择。");
                                return false;
                            }else if (nodes[i].module != "${module}"){
                                top.$.jBox.tip(
                                "不能选择当前栏目以外的栏目模型，请重新选择。");
                                return false;
                            }
                        </#if>
                        ids.push(nodes[i].id);
                        names.push(nodes[i].name);
                        <#if !checked>
                            break; //如果为非复选框选择，则返回第一个选择
                        </#if>
                    }
                    $("#${id}Id").val(ids.join(",").replace(/u_/ig,""));
                    $("#${id}Name").val(names.join(","));
                }
                <#if allowClear>
                    else if (v=="clear"){
                        $("#${id}Id").val("");
                        $("#${id}Name").val("");
                    }
                </#if>
                if(typeof ${id}TreeselectCallBack == 'function'){
                    ${id}TreeselectCallBack(v, h, f);
                }
            }, loaded:function(h){
                $(".jbox-content", top.document).css("overflow-y","hidden");
            }
        });
    });
</script>
</#macro>
```

在该文件中主要定义了两个命令：一个用于菜单的选择，另一个用于菜单图标的选择。它们分别依赖于两个界面并和 TagController 类关联，TagController 类主要用于实现弹窗中界面展示的跳转，其内容如下。

```
@Controller
@RequestMapping("/sys/tag")
public class TagController {

    /**
     * 树结构选择标签 (tagTreeselect.ftl)
     */
    @RequestMapping(value = "/treeselect")
    public String treeselect(HttpServletRequest request, Model model) {
        model.addAttribute("url", request.getParameter("url"));          //树结构数据URL
        model.addAttribute("extId", request.getParameter("extId")); //排除的ID
        model.addAttribute("checked", request.getParameter("checked")); //是否可复选
```

```
        model.addAttribute("selectIds", request.getParameter("selectIds")); //指定默
认选中的ID
        //是否读取全部数据，不进行权限过滤
        model.addAttribute("isAll",
        request.getParameter("isAll"
        //过滤栏目模型（仅针对内容管理系统的Category树）
        model.addAttribute("module", request.getParameter("module"));
        return "sys/tag/tagTreeselect";
    }

    /**
     * 图标选择标签（tagIconselect.ftl）
     */
    @RequestMapping(value = "/iconselect")
    public String iconselect(HttpServletRequest request, Model model) {
        model.addAttribute("value", request.getParameter("value"));
        return "sys/tag/tagIconselect";
    }
}
```

弹窗中的数据展示界面主要由 templates 文件夹的 tag 文件夹中的 tagIconselect.ftl 和 tagTreeselect.ftl 来展示，读者可以自行查阅源码中的数据。

9.2.6 权限控制

基础设施的搭建完成后，将引入 Shiro 来实现权限的细粒度控制。依赖的添加在项目创建时已经完成，这里不再赘述，下面主要讲解具体如何实现。

1. Shiro 权限处理回顾

第 6 章已经对 Shiro 进行了详细介绍，这里只简单回顾 Shiro 的权限控制是如何实现的。

权限控制包括两点：认证和授权。Shiro 也是这样进行权限处理的，即通过认证器对用户信息进行认证，通过授权器进行权限判定。但对于这些信息，Shiro 并不提供维护，而需要由开发人员通过 Realm 提供给 Shiro。

2. 编写自定义 Realm 对象

为 Shiro 提供用户认证信息及授权相关的角色和权限信息时，需要在 config 包中创建 security 子包并创建 TlhShiroRealm 类，其继承自 AuthorizingRealm，内容如下。

```
public class TlhShiroRealm extends AuthorizingRealm {

    @Autowired
    private UserService mUserService;

    @Override
    protected AuthorizationInfo doGetAuthorizationInfo(PrincipalCollection principals) {
        User user = (User) principals.getPrimaryPrincipal();
        SimpleAuthorizationInfo authorizationInfo=new SimpleAuthorizationInfo();
        authorizationInfo.setRoles(
                new HashSet<>(mUserService.findRolesByUserName(user.getUserName())));
        authorizationInfo.setStringPermissions(
                new HashSet<>(mUserService.findMenusByUserName(user.getUserName())));
        return authorizationInfo;
    }

    @Override
```

```
    protected AuthenticationInfo doGetAuthenticationInfo(
            AuthenticationToken token) throws AuthenticationException {
        String username = token.getPrincipal().toString();
        User user = this.mUserService.findUserByUserName(username);
        if (user != null) {
            if (user.getEnabled()) {
                SimpleAuthenticationInfo authenticationInfo
                        = new SimpleAuthenticationInfo(user, user.getPassword(),
                          ByteSource.Util.bytes(user.getSalt()), getName());
                return authenticationInfo;
            } else {
                throw new DisabledAccountException("账户不可用");
            }
        } else {
            throw new AccountException("用户名错误");
        }
    }
}
```

3. 编写 Shiro 的配置类

除了上面定义的 Realm 以外，还要定义 Realm 与 Shiro 核心配置关系的信息，在 security 包中创建 ShiroConfig，并定义 Shiro 的相关配置信息，代码如下。

```
@Configuration
@Import({ShiroWebConfiguration.class,
      ShiroWebFilterConfiguration.class,ShiroAnnotationProcessorConfiguration.class})
public class ShiroConfig {

    @Bean
    public FormAuthenticationFilter formAuthenticationFilter(UserService userService){
        FormAuthenticationFilter    formAuthenticationFilter=new    FormAuthentication
Filter();
        formAuthenticationFilter.setUserService(userService);
        return formAuthenticationFilter;
    }

    @Bean
    public Realm realm(){
       TlhShiroRealm realm=new TlhShiroRealm();
       HashedCredentialsMatcher credentialsMatcher=new HashedCredentialsMatcher();
       credentialsMatcher.setHashIterations(5);
       credentialsMatcher.setHashAlgorithmName(Md5Hash.ALGORITHM_NAME);
       realm.setCredentialsMatcher(credentialsMatcher);
       return realm;
    }

    @Bean
    public ShiroFilterChainDefinition shiroFilterChainDefinition(){
       DefaultShiroFilterChainDefinition shiroFilterChainDefinition
            =new DefaultShiroFilterChainDefinition();
       shiroFilterChainDefinition.addPathDefinition("/login","authc");
       //只要用户登录即可访问该过滤器
       shiroFilterChainDefinition.addPathDefinition("/dashboard/**","user");
       //只要用户登录即可访问该过滤器
       shiroFilterChainDefinition.addPathDefinition("/sys/**","user");
       shiroFilterChainDefinition.addPathDefinition("/logout","anon");
```

```
        return shiroFilterChainDefinition;
    }

    @Bean
    public EventBus eventBus(){
        return new DefaultEventBus();
    }
}
```

在该配置类中,主要对 Realm、Filter 和 Filter 链进行了定义,继续创建 ShiroWebFilterConfiguration 类,其继承自 AbstractShiroWebFilterConfiguration,用于和 Web 容器进行整合。具体代码如下。

```
@Configuration
@ConditionalOnProperty(name = "shiro.web.enabled", matchIfMissing = true)
public class ShiroWebFilterConfiguration extends AbstractShiroWebFilterConfiguration {

    @Autowired
    private FormAuthenticationFilter formAuthenticationFilter;

    @Bean
    @ConditionalOnMissingBean
    @Override
    protected ShiroFilterFactoryBean shiroFilterFactoryBean() {
        ShiroFilterFactoryBean shiroFilterFactoryBean = super.shiroFilterFactoryBean();
        //设置自定义的过滤器
        Map<String,Filter> filters=new HashMap<>();
        filters.put("authc",formAuthenticationFilter);
        shiroFilterFactoryBean.setFilters(filters);
        return shiroFilterFactoryBean;
    }

    @SuppressWarnings("deprecation")
    @Bean(name = "filterShiroFilterRegistrationBean")
    @ConditionalOnMissingBean
    protected FilterRegistrationBean filterShiroFilterRegistrationBean() throws Exception {

        FilterRegistrationBean filterRegistrationBean = new FilterRegistrationBean();
        filterRegistrationBean.setFilter((AbstractShiroFilter) shiroFilterFactoryBean().getObject());
        filterRegistrationBean.setOrder(1);

        return filterRegistrationBean;
    }
}
```

另外,用户登录成功之后会更新用户的最后登录信息,这里采用 Shiro 中的过滤器进行实现。创建 FormAuthenticationFilter 类,其继承 Shiro 原有的 FormAuthenticationFilter 类,并重写 onLoginSuccess 方法,用于监听登录成功之后的回调,代码如下。

```
public class FormAuthenticationFilter extends
org.apache.shiro.web.filter.authc.FormAuthenticationFilter {

    private UserService userService;

    public void setUserService(UserService userService) {
        this.userService = userService;
```

```java
    }

    @Override
    protected boolean onLoginSuccess(
                    AuthenticationToken token, Subject subject,
                    ServletRequest request, ServletResponse response) throws Exception {
        User user = subject.getPrincipals().oneByType(User.class);
        user.setLastLoginTime(new Date());
        userService.updateLastLogin(user);
        return super.onLoginSuccess(token, subject, request, response);
    }
}
```

4. 继承 FreeMarker 的标签库

前面的配置已经实现了通过 Shiro 对权限的控制，但为了实现更细粒度的控制，同时使不同权限的用户登录系统之后看到的界面不同，需要结合相应的标签进行处理。将源码中 FreeMarker 中的类复制到项目对应的 config.freemarker 包中，并在 security 中创建 FreeMarkerConfig 类，将定义的 tags 注入 FreeMarker 配置环境。具体代码如下。

```java
@Configuration
public class FreeMarkerConfig {

    @Autowired
    private freemarker.template.Configuration configuration;

    @PostConstruct
    public void setSharedVariable() {
      try {
        configuration.setSharedVariable("shiro", new ShiroTags());
      } catch (Exception e) {
        e.printStackTrace();
      }
    }
}
```

经过前面的配置，在界面中已可以使用相应的 Shiro 标签来控制按钮的显示，如菜单列表界面中按钮的显示。如下代码所示，通过判断用户是否具有"sys:menu:form"权限而决定是否显示"菜单添加"按钮。

```
<@shiro.hasPermission name="sys:menu:form">
    <li><a href="${ctx}/sys/menu/form">菜单添加</a></li>
</@shiro.hasPermission>
```

9.2.7 项目部署

通过前面的布署开发，系统的所有功能都已经实现，最后要进行打包和发布。打包和发布对于 Spring Boot 项目而言有 JAR 包和 WAR 包两种方式，这里主要讲解使用 JAR 包的方式进行打包并采用内嵌的 Servlet 容器运行项目。

1. 项目打包

修改 pom.xml 文件，添加 Spring Boot 打包所需要的插件，内容如下。

```xml
<build>
    <!-- 设置打包之后的文件名 -->
    <finalName>${artifactId}</finalName>
    <plugins>
```

```xml
        <plugin>
            <groupId>org.springframework.boot</groupId>
            <artifactId>spring-boot-maven-plugin</artifactId>
        </plugin>
        <plugin>
            <groupId>org.apache.maven.plugins</groupId>
            <artifactId>maven-compiler-plugin</artifactId>
            <configuration>
                <encoding>UTF-8</encoding>
                <source>1.8</source>
                <target>1.8</target>
            </configuration>
        </plugin>
    </plugins>
</build>
```

切换到项目目录，执行如下命令。

```
mvn clean package -Dmaven.test.skip=true
```

进入 target 目录，可以看到打包生成的 tlhhup.jar 文件，执行 java –jar tlhhup.jar 命令，可以看到，项目能够正常启动并可通过浏览器进行正常访问，项目启动效果如图 9-9 所示。

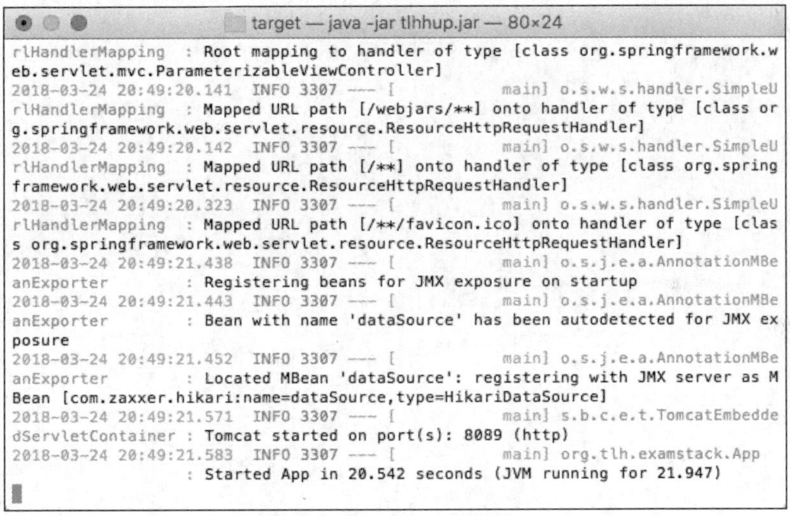

图 9-9　项目启动效果

2．编写脚本

编写 Shell 脚本，以在 Linux 操作系统中进行维护，最终项目将部署在用户目录的 app/tlh/web/tlhhup 目录中，在该目录中创建 web-tlhhup.sh 文件，其内容如下。

```sh
#!/bin/sh
## java env
export JAVA_HOME=/usr/jdk1.8.0_161
export JRE_HOME=$JAVA_HOME/jre

## service name
APP_NAME=tlhhup

SERVICE_DIR=/root/app/tlh/web/$APP_NAME
SERVICE_NAME=tlh-service-$APP_NAME
JAR_NAME=$SERVICE_NAME\.jar
```

```
PID=$SERVICE_NAME\.pid
cd $SERVICE_DIR

case "$1" in
    start)
            nohup $JRE_HOME/bin/java -Xms256m -Xmx512m -jar $JAR_NAME
            >/dev/null 2>&1 &
            echo $! > $SERVICE_DIR/$PID
            echo "=== start $SERVICE_NAME"
            ;;
    stop)
            kill 'cat $SERVICE_DIR/$PID'
            rm -rf $SERVICE_DIR/$PID
            echo "=== stop $SERVICE_NAME
            sleep 5

            P_ID='ps -ef | grep -w "$SERVICE_NAME" | grep -v "grep" | awk '{print $2}''
            if [ "$P_ID" == "" ]; then
                    echo "=== $SERVICE_NAME process not exists or stop success"
            else
                    echo "=== $SERVICE_NAME process pid is:$P_ID"
                    echo "=== begin kill $SERVICE_NAME process, pid is:$P_ID"
                    kill -9 $P_ID
            fi
            ;;
    restart)
            $0 stop
            sleep 2
            $0 start
            echo "=== restart $SERVICE_NAME"
            ;;

    *)
            ## restart
            $0 stop
            sleep 2
            $0 start
            ;;
esac
exit 0
```

3. 添加可执行权限

具体代码如下。

```
chmod +x web-tlhhup.sh
```

4. 进行相关管理

```
//启动服务命令
./web-tlhhup.sh start
//停止服务命令
./web-tlhhup.sh stop
//重启服务命令
./web-tlhhup.sh restart
```

通过本章的学习，相信读者对如何采用 Spring Boot 快速搭建开发环境及整合第三方框架有了更深刻的了解，本章还介绍了如何对项目进行打包以及如何在 Linux 操作系统中进行项目部署，这对程序开发有很大帮助。